7.50

SOIL CONSERVATION IN PERSPECTIVE

SOIL CONSERVATION IN PERSPECTIVE

BY

R. BURNELL HELD

AND

MARION CLAWSON

PUBLISHED FOR RESOURCES FOR THE FUTURE, INC.
BY THE JOHNS HOPKINS PRESS,
BALTIMORE, MARYLAND

RESOURCES FOR THE FUTURE, INC.
1755 Massachusetts Avenue, N.W., Washington, D.C. 20036

Resources for the Future is a non-profit corporation for research and education in the development, conservation, and use of natural resources. It was established in 1952 with the co-operation of the Ford Foundation and its activities since then have been financed by grants from that Foundation. Part of the work of Resources for the Future is carried out by its resident staff, part supported by grants to universities and other non-profit organizations. Unless otherwise stated, interpretations and conclusions in RFF publications are those of the authors; the organization takes responsibility for the selection of significant subjects for study, the competence of the researchers, and their freedom of inquiry.

This book is one of RFF's studies in land use and management, which are directed by Marion Clawson. Burnell Held is an RFF research associate. The manuscript was edited by Virginia D. Parker. The illustrations were drawn or adapted by Clare O'Gorman Ford.

Director of RFF publications, Henry Jarrett; *editor,* Vera W. Dodds; *associate editor,* Nora E. Roots.

PREFACE

FOR more than thirty years, the United States has made a major national effort to encourage soil conservation. Why was the soil conservation program begun? What were its original objectives—what was it expected to accomplish? How much, in fact, has been accomplished since this program was begun? What have been the obstacles to still greater accomplishment? What remains to be done, as the job is defined by the organizations and persons most directly concerned? What broad problems will soil conservation efforts face in the next generation? How might national programs of soil conservation profitably be redirected in the future? These are the questions to which this study was directed and on which we hope it will throw some light.

Our concern, in this book, is primarily with the economics, both individual and group, and with the social and political relationships involved in use and misuse of soil. But we try always to remember that behind these human reactions to soil lies the complex physical structure of the soil itself. The human actions are limited to a major degree by that physical structure, although they in turn affect the soil.

At the outset, however, we must emphasize that a lack of information about soil conservation is much more striking than any of the data which deal with the subject. Time and again in this book, we are forced to point out that needed information is lacking on critical points. For example, available data reveal that 38 per cent of the cropland had been "adequately treated" in 1958. But was this the land most in need of treatment, that least in need, or just average cropland? Depending upon the answer to this question, the total soil conservation job as it appeared thirty years ago can be said to be 60 per cent, or 10 per cent, or 38 per cent accomplished—a wide and nearly meaningless range.

The significance of soil losses lies not in their physical dimension,

v

however, but in their present and future consequences for men. No one would seriously propose that all erosion be eliminated—the undertaking would be too costly for what it would accomplish. The unanswered question is: Where does the effort stop? To this question, there is no single answer. It will be different for every part of the nation. It will depend upon the current and projected uses of the specific land, the erosion hazard associated with various methods of handling the soil in these uses, the effectiveness and cost of alternative erosion control measures and the individual and social adjustments required by each, as well as the consequences of allowing erosion to continue at various levels. These questions have been largely neglected to date.

We have naturally used whatever information we could find, and have tried to point out its meaning. Whenever information is inadequate, there is grave danger of misinterpretation. We have thought it better to run some risks of this, however, than to refrain from analysis and from drawing conclusions. For those who may not agree with what we have said, we have tried to show the basis on which our conclusions rest.

To oversimplify the results of this study, we conclude:

1. Major accomplishments have been achieved by the national soil conservation program.

2. A big job remains to be done, regardless of how that job is defined and measured.

3. Fundamental conflicts exist within programs of the U.S. Department of Agriculture, particularly between its extensive programs aimed at increasing basic land productivity and current output and those aimed at adjusting current output to market needs at desired prices.

4. The broad soil conservation effort faces different and major problems for the next generation, which will call for much greater change in the program than has been incorporated thus far.

A companion book resulting from this same general research project: *Governing Soil Conservation: Thirty Years of the New Decentralization*, by Robert J. Morgan, is being published by The Johns Hopkins Press for Resources for the Future. Mr. Morgan has been concerned primarily with the political and governmental processes by which soil conservation policies and programs came into existence and developed; while his treatment is partly historical, he also analyzes the present situation. His study is rather more detailed than ours and has been heavily documented. We have relied upon his work to a major extent, both in contributing to our general understanding of the problem and, more specifically, in our discussion in Chapters 3 and 4. In our view, the two studies complement one another.

In bringing our study to a close, we must express thanks to a number of persons who have been helpful during the course of this research. First of all, several of our colleagues at Resources for the Future have helped through their interest and by stimulating comments during the research, writing, and editing of this book. Among these is our former colleague, Eleanor E. Hanlon, then on leave from Syracuse University, who made extensive analyses of the National Inventory of Soil and Water Conservation Needs and of various published studies, and whose discussions with us from time to time were most stimulating and helpful. We wish also to acknowledge the assistance of Clifford U. Koh, who assembled most of the statistics used in this book, as well as making other related statistical inquiries.

We also have had a great deal of help from persons in the federal agencies most directly concerned—the Soil Conservation Service, the Production and Marketing Administration and its successor, the Agricultural Stabilization and Conservation Service, the Forest Service, and the Bureau of Land Management. Our greatest debt, as the book will show, is to the considerable army of professional workers of various disciplines, from many public and private agencies, whose published studies we have read, quoted from, or depended upon in other ways. To all of these, we express deep appreciation. We alone, of course, are responsible for the interpretations of the available facts and for any errors that may have slipped past us and the several reviewers of this manuscript.

R. BURNELL HELD
MARION CLAWSON

March, 1965

CONTENTS

PART II.   PERFORMANCE AND EVALUATION

PART III.   THE NEXT GENERATION

List of Figures

List of Tables

PART I

BACKGROUND
AND SETTING

Chapter 1

MAN AND RESOURCES

FROM prehistorical times to the present, man has been concerned with his future food supply. Primitive men learned to store some of the seeds they gathered for the lean seasons and years, and hunters learned to dry meat for future use. A modest store of food provided a reserve which might mean the difference between life and death. The margin between food supply and food need was often close and precarious and the balance had to be struck in a very small area because of the lack of transport. The seven fat and seven lean years of ancient Egypt as related in the Bible became the base for an ever-normal granary several thousand years later. The ancient problem of food supply has become the classic problem of the relation between resources and population, which concerns most of the world today in terms of present needs, and all of it in terms of the long-range future.

Within the period of modern history, this classic problem had its most forceful original statement by Malthus. There were, of course, forerunners of Malthus—Adam Smith, among others—and there were later doctrines, but his statement was so directly to the issue, and so forceful, that it has long stood as the prototype.[1] Malthus took as accepted the facts that resources were limited and that population would multiply continuously unless it was restrained either voluntarily or by various checks such as disease and lack of food. He not only believed there was an eventual physical limit to the quantity of agricultural land, but also that its quality was uniform. And he anticipated no increase in production costs until all land had been employed.

[1] For a full discussion of the variations on the scarcity doctrine developed by Thomas Malthus and David Ricardo in the first quarter of the nineteenth century, and those of John Stuart Mill later, see Harold J. Barnett and Chandler Morse, *Scarcity and Growth: The Economics of Natural Resource Availability* (Baltimore: The Johns Hopkins Press for Resources for the Future, Inc., 1963), pp. 51–71.

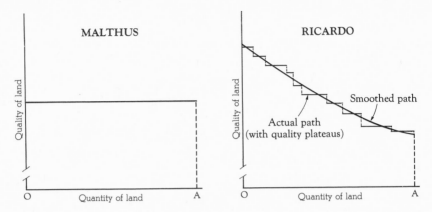

Figure 1. The availability of agricultural land: Malthus vs. Ricardo. (Chart from Harold J. Barnett and Chandler Morse, *Scarcity and Growth: The Economics of Natural Resource Availability,* The Johns Hopkins Press, 1963, p. 60.)

Beyond that point, output could be increased from the same area of land, but only at rising cost, thus introducing (in other terms) the principle of diminishing marginal output.

Ricardo was shortly to introduce variations and changes in the Malthusian model. As shown in Figures 1 and 2, he agreed with Malthus on a limit to the quantity of land, but, in particular, he made much of variations in the quality of land. The second part of Figure 1 puts Ricardo's views on land scarcity and quality of land in modern graphic form. Under his formulation the best land is used first, with increasingly poorer land used as needed. As a result, the cost of output rises long before the final margin of cultivation is reached, as shown in Figure 2. How soon and how sharply costs rise depend upon the area of land of different qualities, on the one hand, and upon the size of the population—hence, upon the demand—on the other.

Malthus, Ricardo, and later writers applied their theories of resource scarcity to other resources than agricultural land. In particular, they applied them to minerals. It was nearly always assumed that the total quantity of mineral was absolutely limited, although differences in grade of deposit were accepted, and that each pound or ton mined meant that much smaller volume remaining for later extraction. Thus, the problem of using such fixed and irreplaceable resources was different from that of using soil, which could be maintained productively forever.

Malthus and Ricardo have left an enormous legacy of conviction in

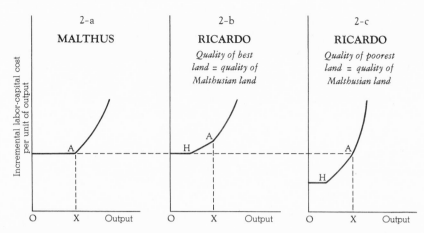

Figure 2. Malthusian and Ricardian economic scarcity effects. (Chart from Harold J. Barnett and Chandler Morse, *Scarcity and Growth: The Economics of Natural Resource Availability*, The Johns Hopkins Press, 1963, p. 61.)

modern conservationists, even among those who never heard of, or never read, either. The idea that men shall some day "run out" of minerals, and perhaps other resources, or that population increase will outrun agricultural output, is deeply ingrained in much popular thinking.

These theories were essentially static in nature. While population increase was basic to the theoretical structure, other change was largely ignored or brushed aside. The possibility of technological advance was recognized, but it was believed that at best this would only postpone the ultimate day of reckoning. According to these views, such advance could not be counted upon to solve the resource problem, much less was it in any way directly related to resource consumption.

In their recent reconsideration of the whole subject, Barnett and Morse make technological change the center of their concern.[2] They view the use of natural resources in a complex modern society and economy as part of a larger process that necessarily includes forces of technological change, which in turn have profound repercussions upon resource use itself. The mining of metals is part of such an economy, but the technological changes arising out of a metal-using economy make the supply of metals that is economically available vastly greater

[2] *Ibid.*

than before. It is true, in a strict physical sense, that each pound or ton of metal mined means one pound or ton less in the ground. However, the total supply in the ground is so vast, if the lowest concentrations are included, as to be nearly inexhaustible. And that which is mined is not destroyed in any absolute sense, but becomes useful economically for other purposes or is ultimately recycled through various natural processes of diffusion and concentration. Under the dynamic theoretical structure erected by these two authors, resource scarcity is no longer inevitable or probable; on the contrary, it is most unlikely.

Not all economists will accept the Barnett-Morse formulation, even for technologically advanced countries; and it may not be applicable to poor countries which lack capital, trained manpower, and technological capacity. Nevertheless, the role of technology in making available resources which otherwise would be unusable is very great, and widely acknowledged. Certainly, we include technological change as one of the major variables in our consideration of soils and their use.

Concepts of Resources

The term "natural resources" has different meanings for different persons. Our definition, stated briefly is: "A natural resource is any quality or characteristic of nature which man knows how to use economically to ends which he desires."[3]

The qualities and characteristics of nature are almost limitless—soils, climate, vegetation, minerals, animals, and many others—and the range within each is very great. Some are important at one stage in human development, while others are useless or even unknown. There is much interest, from time to time, in resource inventories; but any resource inventory is useless except as directed to a specific kind of use and a known technology. The mild climates of Florida, Arizona, and Southern California are extremely valuable assets in a modern industrial economy where attractions to a skilled labor force may far outweigh a local availability of coal and iron, for example.

The role of technology in making qualities or characteristics of nature useful for man's needs is well known. Until men knew how to get petroleum out of the ground, and to use it once it was out, it did not exist in any economic sense. Uranium, once a chemical curiosity, be-

---

[3] A similar view, but not this definition, is found in Eric W. Zimmerman, *World Resources and Industries* (New York: Harper and Brothers, 1933).

came an international strategic material overnight, when the first atomic bomb was exploded; when discoveries so far outran earlier estimates, it became a glut on the market.

But neither technology nor extraction of earth's qualities is costless; many things are possible which are not currently economic. The United States possesses great deposits of oil shale and Canada has great deposits of tar sands. From both it is wholly possible to extract enormous quantities of petroleum, but thus far it has not proved economic to use either. Salt water can be converted to fresh "to make the deserts bloom," but thus far the costs have been prohibitive for any but limited special uses. Power from atomic sources is technically feasible, and seems to be on the verge of becoming economically feasible—at least under some circumstances. Caution is required in speaking of economic feasibility, however, because history shows how much it has changed in the past. Yet, at any given time, certain qualities of nature, such as the oil shale deposits, can be evaluated on the assumption that some day they will become economically usable, even though they are not today. They are part of potential resources, but not part of presently usable ones.

The goals that men seek in resource use are equally important, and sometimes overlooked or inadequately considered. The attainment of the greatest possible income, measured in conventional dollar terms, may be sought for many purposes. This is an appropriate goal, and one for which the tools of economic analysis are well suited. But it is not the only goal, as most economists realize. Resources may be used for some ends which are not easily measured, or measurable at all, in monetary terms. One practical example of considerable quantitative importance today is the use of forests and other land areas for recreation. Others are the use of large quantities of minerals, labor, and capital for defense purposes; and the shipment abroad of wheat and other foods not needed at home to relieve distress and to assist in the economic growth of developing areas. It may be argued that, in each of these cases, the aim is to maximize something—whether personal satisfactions not easily measurable in money terms, or national security, or something else. But, in some cases patterns of resource use seem scarcely to maximize anything, unless it be conformance with long-established prejudices. To stretch the interpretation of income maximization to fit all these cases resembles the efforts of an old hen to spread out and cover an overly large setting of eggs—something is likely to be left in the cold. The problem of goals becomes especially difficult when resource use in many countries and cultures is consi-

dered. What may seem irrational to Americans may be the height of
rationality to the people of the other culture.

These considerations about resources in general apply equally to
soil. And, even in the United States, the objectives of soil conservation
programs are not identical for everyone. Elaborate procedures have
been developed for measuring the physical characteristics of soils—
their depth, geologic history, size of constituent particles, slope, availability of plant nutrients, and many other features. Through practical
experience and scientific research, a rich technology also has been
developed for their use. Soils are used primarily by farmers and other
private landowners, who seek maximum incomes as at least one goal,
and who are thus highly responsive to economic considerations.

Each aspect of our definition of natural resources as applied to soils
has undergone substantial change in the United States over the past
three hundred years. Through erosion, different hydrologic relationships caused by forest clearing and other practices, irrigation, continued cropping, and so on, the basic characteristics of the soil have
been changed in many areas. The enormous advances in technology,
from hand tools in the colonial era through animal power during the
nineteenth century to mechanical power today—and equally dramatic
changes in other aspects of farm technology—have been described
many times. Growth of large cities, development of transportation
networks, creation of marketing institutions, and many other economic
changes have affected the economics of agriculture in still other ways.
The goal of farming has changed to some degree, from provision of
food and fiber for the farm family on a largely self-sufficient base to
the production of commodities for sale in the expectation, or hope,
of a profit.

Resources are so enormously variable and diverse, in so many ways,
that some grouping into broad classes seems essential. From a broad
technological or management viewpoint, resources can be grouped as
shown in Table 1. Some resources are "flow resources" with a stream
of uses possible without loss of the resource itself. These, in turn, are
divisible into nonstorable resources, of which sunlight is a good
example, and storable ones, such as water from natural precipitation.
In contrast are the "fund resources." All of these are exhausted by use,
but some can be renewed and some cannot. The "exhaustible but
renewable" fund resources are typified by mature forests or soil
fertility. Either can be used and renewed, at least under many circumstances. However, this is possible only within a range of reversibility.
If all the trees of a species were cut, to use an extreme example, forests

TABLE 1. General characteristics of natural resources

Type of resource	Relation of use in one period of time to use in following periods	Possibility of increasing supply in later periods by man's present activities	Management alternatives	Examples
I. Flow resources				
a. Nonstorable	None	None, or very little	Use or nonuse	Sunlight
b. Storable	Large, within limits of storage capacity	Considerable	Storage, nonstorage, unstoring	Water stored in reservoirs or in soil
II. Fund resource				
a. Exhaustible but renewable	Large, within range of reversibility	Large, if necessary present steps taken	Maintain, decrease or increase availability over time	Soil fertility, forests
b. Exhaustible and nonrenewable	Complete and inverse; use now exactly reduces use later, and vice versa	Only by refraining from consumption now	Rationing over time, substituting flow or renewable fund resources	Petroleum, peats and mucks

SOURCE: John F. Timmons, *et al.*, Committee on Soil and Water Conservation of the Agricultural Board, *Principles of Resource Conservation Policy, With Some Applications to Soil and Water Resources* (Washington: National Academy of Sciences–National Research Council, 1961), Publication 885, p. 8.

of the same species could not be re-established. Soil can be renewed when nutrients are used and even part of the soil itself washed away or otherwise destroyed; but if the process goes too far it will prove impossible to reverse. Still other resources—the extraction of petroleum and mining of minerals—for example, are "exhaustible and nonrenewable," at least within the span of human planning.

Possibly we should add another class—"nonexhaustible resources." For example, climatic factors, position on the earth's surface, basic geology, and perhaps other aspects or characteristics are so permanent and so difficult or impossible for man to change that they are practically inexhaustible. The same might be said of the minerals dissolved in sea water. However, these can be considered, perhaps without too much violence, as flow resources of a nonstorable kind.

Wantrup uses a system of classification of natural resources (see Table 2) which is similar to the foregoing in some respects, but differs significantly from it in other ways.[4] While his major breakdown is

[4] S. V. Ciriacy-Wantrup, *Resource Conservation—Economics and Policies* (rev. ed.; Berkeley: University of California Press, 1963).

TABLE 2. Wantrup's classification of natural resources

I. NONRENEWABLE OR STOCK RESOURCES.
   1. Stock not significantly affected by natural deterioration: metal ores *in situ;* coal; stones; clays.
   2. Stock significantly affected by natural deterioration: refined metals subject to oxidation; oil and gas in case of seepage and blowoff; plant nutrients subject to leaching; radioactive substances in process of nuclear disintegration; surface water reservoirs subject to evaporation.

II. RENEWABLE OR FLOW RESOURCES.
   1. Flow not significantly affected by human action: solar and other cosmic radiation; tides; winds.
   2. Flow significantly affected by human action.
      a) Reversibility of a decrease in flow not characterized by a critical zone: precipitation; special locations that form the basis of site value; services from a species of durable producer or consumer goods.
      b) Reversibility of a decrease in flow characterized by a critical zone: animal and plant species; scenic resources; storage capacity of groundwater basins.

SOURCE: S. V. Ciriacy-Wantrup, *Resource Conservation—Economics and Policies* (rev. ed.; Berkeley University of California Press, 1963), p. 42.

closely related, his secondary breakdown for the stock resources (or fund resources) is on the basis of absence or presence of natural deterioration, and for the flow resources is inability or ability of human action to affect the resources. Still other classifications are possible; the problem is not simple, and different considerations may loom larger in the minds of some students than in those of others.

In any event, such classifications, while logical and helpful in many ways, generally do not fit any particular resource exactly. Soils, which are our major concern, are complex physical combinations of resources. Some of their characteristics fit into one part of a general classification, others fit elsewhere. Basic position on the earth's surface—with all that this means in terms of geologic origin and base for the soil, climate, and other features—is a stock or fund resource, not naturally changing, not storable, not exhaustible, and hardly subject at all to man's control. Should major control over climate (or any aspect thereof) become practical, this classification would have to be changed accordingly; but that day is not yet. The soil particles themselves, above the basic bedrock, are a fund or stock resource, but are exhaustible; soil can be eroded away, right down to bare rock, and be irreplaceable within human lifetimes. Even short of such drastic destruction, soil can be damaged permanently, perhaps within a restorable range but subject to improvement only at considerable cost. The organic matter contained in the soil, the level of plant nutrients, and even, to an extent, the soil structure are stock in the sense that these properties were part of the soil before it was ever tilled. They are also flow resources in the sense that they change, either naturally or in response to man's use. The changes brought about by man often cause significant shifts in the productivity of the soil, increasing or reducing it. Soil-moisture relationships often can be modified greatly, even though precipitation remains unchanged, by irrigation, drainage, leveling, terracing, and other measures, which may be considered more or less permanent so long as they are maintained.

These considerations applying to soil also apply to forests and natural grazing areas. When Europeans first came to America, the plant cover had many characteristics of a stock resource. There were great stands of mature forests and extensive prairie areas of complex plant associations. These can be materially altered by cutting, fire, and grazing, and were altered, sometimes far beyond the reversible zone, sometimes within it. At the same time, these plant associations had the characteristics of flow resources, in the sense that important quantities of plant materials could be removed annually or at other intervals, and yet leave the basic resource fully capable of rejuvenation.

Productive Capacity, Recurrent Inputs, and Output

Much discussion about soil conservation has been confused because of the failure to distinguish between annual or other periodic inputs for current output and investments made to change the basic productive structure. A manufacturer may vary his output, within limits, without changing the size of his plant or his methods of production, simply by changing the number of man-hours of labor he employs and the quantities of unfinished materials he purchases.

The farmer or land manager has comparable options. He may adjust output in response to changes in costs and prices without changing his "plant" or technology, but by using more or less fertilizer, or some other input. The range of production possibilities open to him under these circumstances is often called a production function. While a farmer may change the quantity of only one input, more typically he will change several inputs at the same time. This may make it harder to determine the contribution of the fertilizer, the improved seed, or the new land management practice, but it is likely to lead to a far more economical production method. The efficiency of more fertilizer may be enhanced if more productive plant varieties are used; or the

improved plant varieties may require heavier fertilization for full results; and heavier fertilization may be even more effective if supplemental irrigation is provided at a critical time in the growing season.

In contrast, the farmer may take other measures in order to affect
the basic productive capacity of the land itself, or for the purpose of
changing the production function or input-output relationship. An
eroding field may be terraced, or a poorly stocked forest planted, or
some other measure taken, with the intent of modifying the productive
capacity of the land. The trend in productive capacity without such
measures may be downward, horizontal, or even upward. Thus, the
measures may be for the purpose of arresting a downward trend, or
for converting a level trend into an upward one, or for accelerating the
rate of an upward trend. In a nation plagued with agricultural surpluses, but concerned over the long-run productive capacity of its land,
there may be good reasons for distinguishing between the maintenance
of present productive capacity in land, and the building of new or
additional capacity—and, in fact, we shall repeatedly make this distinc-
tion in our book. However, in theory, there is little to distinguish
among actions taken to increase basic productive capacity, regardless
of whether the resulting trend is downward, level, or upward, as long
as it is higher than it otherwise would have been.

These relationships are illustrated in Figure 3. In section 3-a of the
figure, the annual input of some factor, such as fertilizer, is varied
with consequent effect upon output. In section 3-b, essentially the
same relationship exists but with productivity at either a higher or a
lower level. Whereas previously the production function was represented by the line PP_1, it is represented by the line CD in the case of a
soil which has retrogressed because of irreversible erosion, or by the
line AB in the case of land where basic productivity has been increased
by drainage, irrigation, leveling, or some other means. The general
similarity of the production functions is evident; but the precise slope
of the lines varies and, more importantly, their level is also different.
Innumerable other lines or production functions are possible.

The relationship between the production function at one time and at
another is shown differently in section 3-c of the figure. Although necessarily only two dimensional, this part illustrates a three-dimensional
relationship. Vertical to the face of the page, or receding into it, are
the production functions shown in the previous two parts. Directly in
the face of the page is the time-productivity relationship involved in
any change in basic soil productivity. In the one case, as shown by the
line PP_t, basic productive capacity remains unchanged; in another

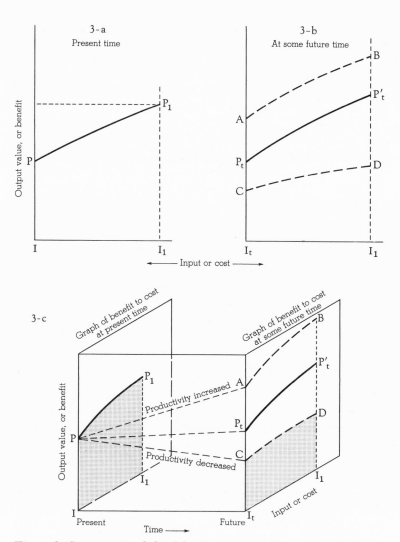

Figure 3. Conservation defined by the relation of investment to benefit at various times, present and future. (Chart from John F. Timmons, *et al., Principles of Resource Conservation Policy, With Some
Applications to Soil and Water Resources,* National Academy of
Sciences–National Research Council, 1961, p. 10.)

case, as shown by the line PA, basic productive capacity has increased considerably over a period of years, while in the third case, as shown by line PC, basic productive capacity has decreased. Some of the reasons for changes in basic productive capacity have been suggested, and others could be cited in particular instances. The possible lines of change are many; some forces might be working to create more productive capacity while others were operative to decrease it. At any level of basic productive capacity, however, a production function or relation between current inputs and current output would exist. During a year or other time period, changes in inputs and in productive capacity might be going on simultaneously.

While these distinctions are most important in theory and in public policy, they are not always easy to apply in practice. A farmer may be faced with a serious soil loss which reduces, and may in time destroy, the productive capacity of his land; he decides to install terraces to cope with this problem. On its face, this may appear as a simple problem in maintenance of basic productive capacity; and, indeed, this result may be achieved. Yet, in order to make the most profitable economic use of the resources now at his command, additional inputs of fertilizer or other materials may be necessary. As an economically rational consequence of the decision to maintain basic productive capacity, therefore, the new level of inputs is higher than the old one. But it also may turn out that the terrace which stopped the soil loss also changed the soil-moisture relationships so as to produce a greater response than in the past from the same plant varieties and the same fertilizer. Productive capacity thus is not merely maintained, but also increased; a new production function comes into existence. Many other illustrations of these complex interrelations could be cited, and an examination of the record for the past thirty years or more suggests that a combination of results has been more typical than a single one. That is, a considerable proportion of the measures to increase current output have significantly increased basic productive capacity, and most measures to maintain productive capacity also have actually helped to increase it. Theoretically, one could undertake just that amount of some remedial measure which would keep productive capacity exactly constant; in practice, the curing of some soil deterioration often opens the way to increased output through increased productive capacity.

In any case, land cannot be considered alone. Land is productive only if combined with labor, capital, current production materials, and management—and management is often the scarcest and most limiting factor of all. To change either current inputs or basic productive capacity requires a managerial decision and new practices. What seems easy to the outsider may seem difficult or impossible to the farmer or land manager. In economic analysis, it is common to show variations in combinations of productive factors—different amounts of labor or capital or both with a fixed area of land, or different areas of land with a given labor supply, and so on. But to many farmers, these factors are fixed. First of all is his own labor and management, both in quantity and quality. Often, he is not interested in a program which would leave him totally or partially unemployed on the farm, or which he feels is beyond his ability. He may be unable to borrow money at any interest rate, or at a rate which makes an investment profitable; or he may be opposed to debt on any terms, regardless of its estimated effect upon his net income. From time to time, we shall come back to the matter of what is realistic for the farmer, given the situation in which he actually finds himself—not in the ideal situation of readily adjustable factors of production.

John F. Timmons and other members of the Committee on Soil and Water Conservation of the Agricultural Board, advance the following definition of conservation: "An investment (1) in maintaining productive potential, (2) in decreasing the productivity deterioration or (3) in enhancing the productivity potential."[5] This is a useful definition, although it might be objected that sometimes measures other than investment are undertaken for the same purpose and have the same effect. However, we accept the emphasis upon changes in basic productive capacity from what would otherwise exist or come into existence. Despite such definitions, "conservation" has come to have so many meanings in popular usage that the word does not have one precise and specialized meaning which everyone will accept. Accordingly, we often shall refer to changes in annual inputs and to changes in basic productive capacity rather than use the word conservation except when referring to special programs known by that term.

Soil Conservation and Productive Capacity

Any program of soil conservation is likely to involve intertemporal, interspace, and interpersonal comparisons—that is, relationships among time elements, geographical areas, and individuals—as well as differ-

[5] John F. Timmons, *et al.*, *Principles of Resource Conservation Policy, With Some Applications to Soil and Water Resources* (Washington: National Academy of Sciences–National Research Council, 1961), publication 885, p. 4.

ence in levels and trends in output. Each of these relationships must be looked at with some care.

If soil conservation is defined as efforts to change the trend in basic productive capacity upward from what it would otherwise be (including holding it constant when otherwise its trend would be downward), then conservation means a different distribution through time of annual inputs and outputs than would otherwise exist. Some writers seem to say that the situation is always one of current investment to be measured against a future stream of increased returns, but this is an oversimplification of a much more complicated situation.[6] The complexity is demonstrated by Barlowe's consideration of different time patterns.[7] These are shown in Figure 4. In the first case (section 4-a of the figure), adoption of a conservation practice or practices prevents a further downward trend in basic productive capacity without any loss in present income. Barlowe suggests that application of lime and fertilizer or adoption of strip cropping or summer fallowing may be in this category. In the case of sheep grazing on native range, where the sheep are herded in any case, a program which deferred some range each year until after seed formation and dispersal might also fall in this category. Variations in methods of harvesting timber, but with a cut of the same volume and value and at no greater cost, might leave the land in much more productive condition. The essential ingredients here are knowledge of the improved practice and a desire to practice conservation—each of which has often been lacking.

Section 4-b shows a case where some immediate sacrifice in income is necessary in order to return future basic productive capacity to the present level, when otherwise it would continue to decline. Barlowe suggests that the building of terraces and check dams, or a shift from row crops to forage crops, pasture, and livestock, may often be in this category. For range lands, a period of recuperation by reduced grazing or even complete rest would also be an example; so would reseeding of the range, except that this would often bring basic productive capacity above the present level. A program to defer the timber harvest for some years, in order to allow trees to get larger and perhaps to grow faster, would also be in this category. Since this is the typical case

----

[6] See Ciriacy-Wantrup, *op. cit.;* also Arthur C. Bunce, *Economics of Soil Conservation* (Ames: Iowa State College Press, 1945); and Anthony Scott, *Natural Resources: The Economics of Conservation* (Toronto: University of Toronto Press, 1955).

[7] Raleigh Barlowe, *Land Resource Economics—The Political Economy of Rural and Urban Land Resource Use* (Englewood Cliffs, N.J.: Prentice-Hall, Inc., 1958).

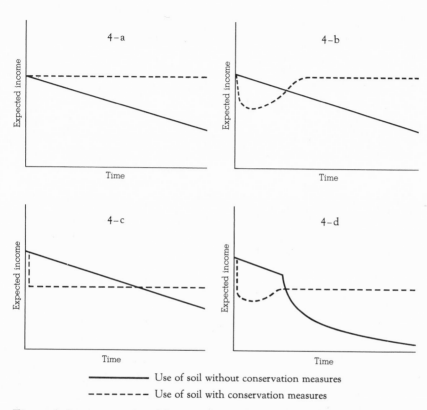

Figure 4. Incomes expected from soil resources over a period of years with and without conservation measures. (Chart from Raleigh Barlowe, *Land Resource Economics: The Political Economy of Rural and Urban Land Resource Use,* © 1958, by permission of Prentice-Hall, Inc., Englewood Cliffs, N.J., p. 300.)

discussed by most economists, we shall—as Barlowe does—explore it in a little more detail after briefly considering the other cases Barlowe uses.

There are small prospects in the case of section 4-c that any program will continue future output at the present level and still maintain basic productive capacity. The operator has two choices: Accept a smaller return now and indefinitely into the future, but with a level trend in capacity. Or experience a continued decline in productive capacity. Note that he does not have the alternative of continuing the present method of operation until output falls below the present long-term sustained capacity, and then shifting to a conservation program at the same level. By that time, the basic productive capacity will have declined further, and he still must make an adjustment downward from his then level of operations. Perhaps there is no program of cropping which is suited to the natural conditions of the area, and a shift to grass or trees is necessary. Or perhaps the area has been persistently overgrazed and the only remedy is fewer livestock. And still other examples might be suggested.

The farmer, in the fourth example (section 4-d), is faced with a moderate decline in productive capacity for some years, after which a crisis will be reached and productive capacity will take a nosedive. This situation might come about if sheet erosion were gradually taking away rather shallow soil materials which were above bedrock or gravel or sterile subsoil. At some point, continued crop production would be impossible. An immediate drastic change may permit continued operations at a lower level than at present but still at a steady rate. The same situation could exist in either range or forest lands if one or more productive plant species were threatened with extinction.

Barlowe's more detailed consideration of the second situation shown in section 4-b takes into account the effect of different discount rates upon future income (see Figure 5). The expected future net returns with no conservation program, shown as ERN_1, have a continuous downward trend. The expected annual net returns from the conservation program are shown as ERC_1. At first, they dip sharply, but come back to the former level in a few years and remain there. Annual returns from the conservation program equal annual returns from the soil-depleting program by time period t_1; but a deficit of net income has accumulated in the intervening years. If the operator used a zero rate of discount of future income—that is, if he prized income to be received in the relatively distant future as highly as income in the near future—then it would be advantageous to adopt the conservation practices so long as his planning period ran well beyond t_1, for the gains after that point would far more than offset the losses up to it.

If the farmer applies the same discount rate to future income from the program without conservation and that with conservation, then the present worth of expected future incomes drops to ERN_2 and ERC_2, respectively. The crossover point, where annual returns from conservation are as high as without it, is at the same time period as formerly, t_1. But the expected future surplus is lower and thus will offset the earlier deficit only after a longer interval; and the higher the discount rate, the longer the time period required to overcome the

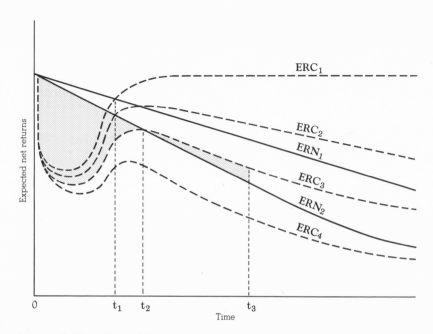

Figure 5. The effect of discounting expected net returns from soil conserva-
tion compared with continued exploitive soil use. (Chart from Raleigh
Barlowe, *Land Resource Economics: The Political Economy of Rural and
Urban Land Resource Use,* © 1958, by permission of Prentice-Hall, Inc.,
Englewood Cliffs, N.J., p. 301.)

initial loss. Barlowe suggests reasonably that the farmer may apply a
higher discount rate to expected income from a conservation program,
partly because of uncertainty on his part as to its actual effects. If so,
the discounted future income stream might look like ERC_3, where a
very long period of time is required for expected future gains to offset early losses, or like ERC_4 where gains never would do so.

This analysis is made entirely in terms of the expected annual net
income to the farmer or landowner, and is useful in these terms. However, it does not take account of the possibility of values in the market
place capitalizing the differences in income streams. If the land market
were perfect—that is, if all future changes in income were perfectly
known and adequately discounted to the present—then future differences in the income stream would be reflected fully in the present land
value. There would still be a problem of the most appropriate interest
or discount rate, and this might well vary between individuals. But the

present value of the farm in the situation shown in section d of Figure 4 would be sharply depressed under present practices. If a conservation program for any of the situations shown in Figure 4 had a favorable balance in terms of present net worth, at the appropriate discount rate, then the present market price would reflect that favorable balance. True, the farmer could capture the difference in land value only by selling the land, but the market price would make clear to him the advantages of the practice.

However, as nearly as we can tell from the imperfect data, which we examine in Chapter 10, the land market seems largely to ignore farm-to-farm differences resulting from conservation programs and trends in basic soil productivity. This partly reflects the fact that soil conservation still is both imperfectly understood and imperfectly adopted. However, it perhaps also reflects the fact that a conservation program is controlled by the farm operator, and may be abandoned or modified by him. A farm with a good conservation program might be sold and a drastically different program adopted; or a run-down farm might be bought by someone who would build it up.

Furthermore, as we have said, soil conservation almost always involves comparisons among areas. The soil material washed or blown away from one spot always accumulates somewhere else. The point of accumulation may be down the slope in the same field or elsewhere on the same farm; in the channel of the small stream of the locality, or in a major river on its way to the sea; or it may be in the sea itself. There are circumstances under which the deposited material may be a benefit—for example, the continued fertility of the Nile Valley over the millenia has been attributed to the annual deposit of fine sediments brought down by flood waters. But, more commonly, the material deposited by wind or water damages the place of arrival as much as its loss does the place of origin. In addition, a changed streamflow resulting from erosion and deposition often does material damage through more severe floods or flood waters containing more sedimentary materials. The distance between the areas of where the soil materials originated and those where they are deposited or causing enhanced problems may be relatively short or relatively long.

Efforts to combat soil erosion frequently lead to disputes among the persons concerned. Generally speaking, at least some of the benefits of a soil conservation program carried out by one landowner are passed on to other landowners. Ownership and occupancy boundaries of land rarely coincide completely with natural boundaries containing erosion, deposition, and changed water relationships. It is possible for a farmer to control gullies or carry out other soil conservation programs and gain the complete and total benefit from them. More commonly, however, some of the benefits accrue farther down in the watershed, perhaps to urban dwellers many miles away. There has been a good deal of exaggeration as to the relative values of on-site and off-site benefits, but the latter are real in many circumstances. They are received by other landowners, or society as a whole, not by the owner of the land on which the erosion occurs or where the conservation program is undertaken. Certainly, the individual landowner cannot be expected to take into account general social benefits, unless he gets cost-sharing payments or unless he is compelled by law to take such measures.

Another, less tangible, interpersonal relationship arises from soil conservation programs. If conservation is widely undertaken, especially under the stimulus of national financial programs and other programs aimed to encourage them, the effect upon total agricultural output may be considerable. In turn, this will affect prices received by farmers, whether they engage in conservation programs or not, and may affect the scale of the national effort required to maintain higher prices for agricultural commodities.

Finally, there is the fact that many soil conservation programs affect the level of output as well as the trend in basic productive capacity. That is, while it is theoretically possible to design a soil conservation program which exactly maintains basic productive capacity and annual output at present levels, in practice, many programs result in an upward trend or higher level of capacity and an enlarged annual output, often achieved by means of a greater annual input.

Attributes and Properties of Soil Resources

Soils are complex mixtures of physical and biological materials that serve a number of functions in agriculture and in other human activities.[8] A natural soil, undisturbed by man, reflects five basic factors: its parent material, climate, internal biological activity, relief of its surface and the surrounding area, and time. Each factor varies widely, and all of the factors interrelate. Each affects or modifies the

[8] For general and authoritative, but not excessively technical, discussion of soils and their properties, see U.S. Department of Agriculture, *Soils and Men—The Yearbook of Agriculture, 1938* and *Soil—The Yearbook of Agriculture, 1957* (Washington: Government Printing Office), and the references cited in them.

effect of the others on the soil. With many variations in each of five factors, it is obvious that a great many theoretical or mathematical combinations are possible. In practice, some of the possible combinations do not exist—for example, a rich biological fauna and flora is impossible in a soil receiving very little precipitation. Nevertheless, soil scientists have distinguished hundreds of different soils in the United States which vary in one or more significant respects from other soils. These varied soils are not sharply and clearly set aside from each other, but are separated by transition zones and gradations. The differences among soils are of crucial importance, not only because of their many consequences for agriculture, but also for other activities based upon soil.

The soil performs numerous functions. It is a physical medium in which plants can put down roots, and thus hold themselves up to light. While this may seem an elemental and perhaps unimportant function, some nearly pure sands which do little more than this can be made productive for various purposes. The soil holds moisture, which the plants can draw upon. Actually, a soil consists of millions of tiny reservoirs, into which water flows when rain or other precipitation occurs or when the land is irrigated or overflowed, and from which the plants draw water as they need it. This reservoir function is extremely important; without it few plants could survive in most climates.

Soils naturally have fertility. That is, prior to their use by man for cultivated crops or for managed grazing or forestry, the soils have fertility which supports the native vegetation or will support crops when used for this purpose. This natural fertility varies greatly, from very fertile to nearly sterile soils. Plants can use natural fertility as it is released slowly from the rocks, soil particles, and other material in which it is held; if fertility were released rapidly, it would be flushed out of the soil by movements of soil moisture, and thus be gone before plants could use it effectively. Soil also can hold and release added soil fertility or soil amendments such as lime. However, soils differ greatly in their need for added materials, and in their capacity to use them efficiently for agriculture, forestry, or grazing. The added materials may be combined chemically with materials already in the soil and released too slowly to be of much benefit to plants; they may be washed out of the soil too easily and rapidly; they may not be necessary because the soil already has enough of the particular chemical element. Soil is a home for microbiological fauna and flora, which in turn play an extremely important role in releasing chemicals from parent material and in other chemical transformations. The soils absorb humus from dead plant materials of various kinds; this then provides a source of food for plants above ground and for the microbiology within the soil, and also helps to transform plant materials in the soil into more readily available chemical forms. In fact, one may accurately view the soil as a vast chemical laboratory, in which many different chemical processes are proceeding simultaneously and with multiple effects.

Soils can be altered in many ways. Although one can imagine a soil unaffected by man, few, if any, were ever completely isolated from all human actions. Even under such an unrealistic primitive state, fire caused by lightning or other natural causes would still be a factor. And primitive man used fire, often consciously, to affect vegetation, which in turn affected the soil. The numerous effects which modern man exerts upon soil may be divided into three general categories: (1) permanent impairment, (2) temporary damage, and (3) improved productivity.

Soils can be more or less permanently downgraded from the point of view of economic use. Soil materials can be physically removed from their original site by water, wind, and man's earth-moving equipment. Man can affect the volume or speed of water as it flows over land, or he can alter the vegetative cover which protects soil against the effects of wind and water. The seriousness of physical loss of topsoil depends in large degree upon what lies beneath it. If the topsoil is underlain with similar soil materials, then its loss, while serious in some senses, still does not prevent the possibility of the development of new topsoils which may be productive. If the topsoil is shallow and overlays solid bedrock, coarse sands or gravel, or infertile clays, however, the loss of the topsoil means that new soils can be developed only over long periods of time—long by geological standards, and never from the point of view of persons now farming the land. In terms of the analysis in Figure 3, the production function has been shifted downward, when soil productivity is permanently impaired, so that the same inputs will no longer produce the same output as formerly. Or, in the terms of our earlier definition of natural resources, the qualities of nature have been changed, without implying anything as to changes in technology, economics, or goals of resource use.

Soils can be temporarily depleted of fertility or otherwise impaired for crop production, but within a rather readily reversible range. Not only may the supply of available plant nutrients be depleted, but the microbiological life of the soil may be reduced in various ways with consequent adverse effects. The internal structure of the soil may be damaged, to take an extreme example, by puddling, which renders it nearly impervious to the movement of water. Soil-moisture relationships may be impaired in such ways as impairing natural drainage or creating excessively rapid internal soil drainage through the cutting of new drainage channels. Alkalinity problems may be created in irrigated areas if care is not taken in managing the soils receiving the water. However, as long as the reduction in productive capacity is not permanent, the former production function can be restored.

Soils can be improved for cultivated crop production in one or more of several important ways. They can be cleared of stones in the plow zone; they can be leveled or smoothed; they can be irrigated; and they can be drained by surface open drains or by underground tile. Fertility can be added by use of chemical fertilizers or natural manures; and soil acidity can be adjusted to crop needs by the use of lime or limestone. Through cultivation, especially by the turning under of plant residues, the soil's structure and its tilth can sometimes be improved materially over the natural condition. Relatively barren desert soils or relatively sterile sands can be made into productive soils by combinations of these processes. Soil change is not invariably a process of deterioration; some soils now in cultivation are more productive for agricultural crops than they ever were in their natural state. If one places such changes in the framework of analysis set forth in Figure 3, these improvements in soil productivity, assumed to be more or less permanent, raise the production function to a new level. If the increase in annual output results from greater inputs—more fertilizer, for example—then the change is primarily a shift along a given production function.

Although the foregoing discussion is all in terms of soil, it applies generally to the natural or native vegetation which was on the land or would become re-established if given a chance. Forests can be reduced in productivity by improper cutting; this may proceed so far that the productivity of the area is permanently reduced, either because soil erosion follows or because previous intricate and delicate ecological relationships cannot be re-established within time periods relevant for individual planning. The productivity of natural grazing areas also can be temporarily or permanently reduced by too heavy or improper seasonal grazing, or in other ways. But nearly all forest and grazing areas can be used on a continuous sustained-yield basis with proper management practices, and sometimes their productivity in physical terms can be stepped up by means of suitable management practices.

Their economic productivity can always be increased above the natural state, for in that state they produce little output.

Motivations for Soil Conservation

Soil is still critical to agriculture, grazing, forestry, and other activities, in spite of all the technological progress of recent decades. It is scientifically possible that some day algae or ordinary agricultural crops may be grown in solutions or in other ways so that all the needs for food, fiber, and wood may be produced without the use of soil. But, to date, such efforts have been much too expensive except under very unusual circumstances. Plant production from the soil is still by all odds the most efficient way to use solar energy. There is much talk, quite properly, about how solar energy is likely in some relatively distant future to replace atomic energy—which in the meantime may have replaced, or at least supplemented, energy from fossil fuels. But man has long used solar energy; we have called it agriculture and forestry. Plants—those which evolved without man's direction and those which have evolved under his management—convert solar energy into many usable materials more efficiently than any other means yet developed for use of this energy source.

Although soil is basic, and likely to remain so for many generations if not forever, man's dependence on soil as such has declined in recent decades and is likely to decline further in the decades ahead. Agricultural output is always the result of labor and capital mixed with land, and in recent years, as we shall show in more detail in Chapter 5, the input mix has changed. Although the hours of labor input have declined by half or more over earlier times, the quality has improved greatly; capital and annual production inputs have also changed greatly in form and have also increased considerably. More important, the input-output relationships—or production function—has changed substantially, so that output has doubled while total inputs have increased only modestly. In this picture, the relative role of land has changed. While land is still basic, other inputs have to some extent reduced reliance upon land; they can substitute for land, within some limits. It seems probable that further changes in this direction will take place during the next few decades.

One major motive for soil conservation is increased income. Because of the differing time distribution of inputs and outputs which conservation nearly always entails, the relevant comparisons must always be in terms of the present worth of the future streams of costs and returns. This, in turn, necessarily involves an interest or discount rate, with many difficult problems of choosing the proper rate. The balance between present values of future inputs and future outputs measures profitability of the proposed conservation measure or program, and often this will be highly influential, if not decisive. Obviously, the operator must consider the costs he will incur and the benefits he will receive.

But, in practice, comparison of present and future costs and returns is not simple. For one reason, each decision maker is faced with major uncertainties. Just how will the proposed measure or program affect output? Professional workers in agriculture and other land-using occupations must admit, in honesty, that often they do not know exactly—they may have an idea, especially as to direction and perhaps as to relative magnitude, but careful calculations of future benefits require a degree of precision that is frequently lacking. Moreover, the man who makes the decision wants to know exactly what will happen over the years ahead if he does not adopt the proposed conservation practice or program; and, again, the professionals often cannot tell him exactly. Moreover, the profitability of the proposed practice or program may depend to a large degree upon the price trends of various agricultural commodities, a matter on which there is always much uncertainty.

Another complicating factor for the decision maker is the rate of interest or discount he should use for future costs and benefits. For many farmers, the interest rate is a subconscious matter, rather than one of careful estimation and calculation; but his preference for present income, as compared with future income, is real enough, even when he cannot be precise about the rate to use. The most careful calculations, which also would seem rational to other decision makers, may justify interest or discount rates as high as 10 per cent or as low as 2 per cent, at a time when the market rate—however defined—is 5 per cent. A farmer short of capital—and "capital rationing" is widespread in agriculture, especially among low-income farmers—might well be able to use his limited capital and credit to produce a high rate of return in his current farm operations. And although he may be able to borrow money at lower cost, like many farmers, he is unwilling to incur debt, either because of the risk involved or because of his general aversion to debt. In a different situation, a landowner with a high personal income, whose marginal income tax rate is 50 per cent or more, might well invest in land—especially if he has hopes of achieving a capital gain thereby—at an effective interest rate of no

more than 2 per cent. There is reason to believe that much speculative
holding of suburban land for a future rise in land values operates on
interest rates as low as this.

Closely related to a decision of the most appropriate interest rate
is that of the relevant time horizon for planning. If the land market
accurately reflected conservation investments, then the length of life
of the decision maker would not be limiting. However, as already
suggested, the land market seems to reflect poorly, or not at all, the
differences in soil conservation of individual farms or properties. Instead, the decision maker must contrast streams of future costs and
future income for a time period which is relevant specifically to him.
This may be as long as 30 years or more; but for many farmers it may
be 20, 10, 5, or fewer years. Especially for an older farmer on a relatively poor farm, the time horizon may be very short; and in all
honesty we cannot urge him to make much sacrifice of current income
for hoped-for future income gains. His current income may be so low
that the present value of a dollar available to him a few years hence
will be heavily discounted. He might not live to receive the dollar, nor
might his heirs benefit either.

The effect of interest rate and time horizon is very great. An investment now which cost $1,000 must result in a very high increased
annual net income if the time horizon is no more than five years, and if
the interest rate is 5 per cent or more. Lengthening the time horizon
to ten years, or lowering the interest rate, will make much less productive investments profitable over the whole period.

Still another complication for many decision makers, faced with soil
conservation problems, is the fixity of inputs. The typical operator of a
small farm has his own labor to sell; operating the farm may be the
only way he can sell that labor. The marginal cost of labor to him is
zero, or nearly so. He will not be interested in any plan for retirement
of all, or much, of his land unless he is compensated for doing it. His
capital input is likely to be limited; so is his land input, since he will
often lack the means to buy or rent additional land.

All of this leads to the conclusion that much soil conservation in
the United States, or the lack of it, is not based solely on profitability
or its lack; and some is not even much concerned with profitability.
Economists and other social scientists may argue that profit considerations should be more influential than they are; the methods of economic analysis, when income maximization is the goal, are powerful
and useful. But some individuals will carry out conservation measures
simply because they like to see well-kept fields and forests, without

too much regard for profitability. Society does some of this also; the tests of economic rationality have never been applied to the control of forest fires, for example, partly because the sight of burned-over forests is offensive. Some landowners do not practice the soil conservation that others would consider profitable; this may be because they are ignorant, indifferent, or because for them the profitability is unproved or absent. However, the tests of economic efficiency for whatever soil conservation programs can still be applied. Is the proposed method the cheapest way to achieve the desired results?

Chapter 2

THE CONSERVATION MOVEMENT,
1890-1920

THE ambiguities that have become imbedded in the word "conservation" extend to a discussion of conservation movements. However defined, conservation—or soil conservation as a specialized branch of the larger field—cannot be considered solely as an economic matter, as a balancing of costs and returns now or in the future. Conservation is a deeply felt social, political, and ideological goal; moreover, it cannot be understood without reference to other social, economic, and political issues and movements of the same time.

The first national all-embracing conservation movement in the United States, which we describe in this chapter, has sometimes been called the "first conservation movement." But this terminology, as applied to the 1890–1920 movement, does not mean there were no earlier ideas or efforts on behalf of conservation. In fact, conservation as a broad issue has concerned some men throughout history; and there have been later conservation movements—notably the soil conservation movement which began in the early thirties. As a social movement, however, the drive which ended roughly in 1920 is more or less accurately designated as the first conservation movement, and there is a very extensive literature about it and the persons involved in it, persons who were also deeply committed to other issues of the time.[1]

---

[1] We can only touch briefly here on the voluminous accounts of the period. For a brief but inclusive review of this first conservation movement, especially from the viewpoint of the social sciences, see Harold J. Barnett and Chandler Morse, *Scarcity and Growth: The Economics of Natural Resource Availability* (Baltimore: The Johns Hopkins Press for Resources for the Future, Inc. 1963), and the references cited therein; see also Henry Jarrett, ed., *Perspectives on Conservation: Essays on America's Natural Resources* (Baltimore: The Johns Hopkins Press for RFF, Inc., 1958), especially the essay by Ernest S. Griffith.

That first conservation movement was a political and social crusade, primarily concerned with natural resources—if the term "natural resources" is very broadly defined. Monetary valuations of natural resources were largely rejected in favor of a less clearly defined, but more deeply held, conviction as to their ultimate worth to mankind. Waste and exploitation were abhorred, thrift and saving of resources encouraged. The movement included a great propaganda campaign to make people aware of the menace of resource destruction, and it also put forward many specific public programs aimed at broad objectives. It was more a mass political and social movement than a clearly thought-out, intellectually derived economic program. From an original concern with natural resources, many of the movement's leaders took most of its followers into activities concerned with immigration, industrialization, trust-busting, pure food, child labor, and other issues of the day. (Not all of the persons concerned with these side issues were involved in conservation, however.) The breadth and lack of sharp outlines of the conservation movement almost surely contributed to its political strength, but they make it difficult, even from this vantage point in time, to describe in sharp intellectual terms.

Our concern is not with the first conservation movement as such, but as an antecedent to the later soil conservation program. Nevertheless, an understanding of how the general conservation movement arose and evolved, and how much the soil conservation movement owed to its forerunner, is needed in order adequately to understand soil conservation.

Political and Socioeconomic Background

The political and socioeconomic background and origins of the 1890–1920 conservation movement go back to the first colonists in North America. From the time of earliest colonial settlement, there had been a plentiful supply of the then known natural resources—especially of forests and the soil. In fact, for much of that period in many parts of the continent, forests which today would be considered wonderful were considered a curse, something to get rid of so that crops could be grown on the land. Magnificent virgin stands of timber were cut and the trees burned in many areas. With such abundant supplies, the known resources obviously could not have a high value compared either to scarce labor or still scarcer capital. The low value and ready availability encouraged liberal use of resources; there was little eco-

nomic incentive to preserve basic productive capacity when this conflicted with exploitation for immediate profit, and still less incentive to
invest for its future increase. Also, technical knowledge on how to
conserve resources often was lacking. The forests of the day were so
extensive, relative to available means of transport, that they seemed
literally limitless to the foot traveler. Land was cleared, farmed, and
abandoned, often within a generation.

Another aspect of most of the nineteenth century and earlier days
in the United States was the ready availability of public land. While
the terms on which such land was available were not as generous by
contemporary standards as they now seem, the statistics on land area
transferred from public to private ownership make it clear that land
was easily available to many thousands of persons. The disposition of
the public lands was a headlong, often heedless, process; it was accompanied by much fraud, speculation, and waste—waste in terms of lives
and money spent attempting to develop land of too limited capacity
for successful farming or other commercial use. The public land policy
can be defended as a necessary part of the process of building a major
new nation on a previously empty continent. But, however one strikes
a balance on the whole period, it certainly inculcated an attitude of
lavish or wasteful use of resources.

The conservation movement owes a great deal intellectually to the
work of Charles Darwin and George P. Marsh. Darwin's influence was
only indirect; he did not espouse conservation as such, but emphasized
the struggle for food and the laws of nature in general. And he greatly
spurred the development of the scientific method, especially as it
applied to nature. Marsh's contribution was more direct; his 500-page
book, *Man and Nature,* published in 1865, was a pioneering inquiry
into the relation between man and his environment.[2] He traced man's
dependence upon nature for food and other necessities of life and, in
turn, described man's ability to modify that environment. There were
other scientists of the day who also made significant contributions.
Coming at a period when human knowledge was beginning to flower
in a major way, but when little attention had been paid specifically
to the natural environment, these men did much to open up new
vistas of fact and ideas for the more ardent and action-minded reformers who were to follow them.

Chief among these later reformers was Gifford Pinchot. He not only
made great contributions himself, but he attracted and helped develop

[2] George P. Marsh, *Man and Nature; or Physical Geography as Modified by
Human Action* (New York: Charles Scribner, 1865).

most of the other active conservationists of the day. His achievements were many: in propaganda or education, to arouse the citizenry; in science and knowledge, to show what could be done in certain specific fields; in resource administration, to help get the national forests set up, and later to administer them; and in politics, to rally an enthusiastic and energetic President to his cause.[3] It is hard to evaluate the specific results of the conservation movement because of their complexity and variety, but there is no question that the movement, gradually developing from this general background, gathered strength and achieved results as it progressed.

Concepts, Objectives, Methods, Conflicts, and Results

In attempting to describe a mass social movement with such a multifaceted appeal as the conservation movement, one must often infer the concepts and meanings of its chief intellectual architects rather than merely summarize their writings. Since Barnett and Morse have done this, we draw upon their presentation in our review of the period.[4]

The conservation movement, Barnett and Morse point out, "abounds with estimates and descriptions of physical natural resources and exhortations to amplify these estimates." The conservation leaders were firmly convinced of the absolute physical limits of supplies of natural resources, and were much concerned that in some cases there would be an early exhaustion of available supplies. Pinchot believed that the nation was on the verge of a timber famine; he preached, and some followed him, that timber prices would shortly rise to nearly prohibitive levels—and investments made on this basis were shortly to prove seriously wrong.

The conservation movement strongly emphasized "ecological balance," and offered many useful insights into the complex interrelations among various natural resources. Its leaders were disturbed about ecological damage—permanent changes in the environment which would leave man poorer for future generations. Changes induced by drainage, cutting of forests, plowing of grassland, and other activities of modern life were believed to have far-reaching and seriously adverse effects. They were particularly disturbed by mineral and fuel depletion.

[3] Pinchot wrote other books, but in *Breaking New Ground* (New York: Harcourt, Brace and Co., 1947), his autobiography, his aims, methods, and results are described in his own—by no means modest—terms.

[4] *Op. cit.*; see especially pp. 75–86.

The waste of natural resources was a major concern of the conservation movement. Waste, in the view of the conservationists, could take one or more of several forms. One of these was the destructive use of resources. Use of fuels to generate electricity when hydropower was unused was a form of waste; so was pollution of streams, plowing of land better suited to grazing, or any one of numerous other changes in natural environment. But underuse of renewable resources, such as timber, was equally wasteful in the view of this group. So was mismanagement of nonrenewable resources, such as fires in underground coal mines, flooding of mines before all the ore was extracted, and the like. Also, the wrong use of extractive or final products, such as flaring of gas for which there was no market, or failing to recover as much secondary and scrap metals as possible, was waste. In the more extreme literature of the movement, waste was nearly equated with sin—a moral, not merely an economic, issue.

The conservation leaders also worried over the social and economic effects of resource scarcity—the scarcity which they believed was either at hand or imminent. There was great glorification of agriculture, the farmer, the simple life; the city often was equated with squalor, disease, and immorality; and there was fear of bigness—whether in industry or in labor unions. Resource scarcity was expected to lead to monopoly, and in turn it would be fostered or accelerated by the monopoly it helped to create. With growing resource scarcity, prices of raw materials were expected to rise, inhibiting economic growth and well-being.

This listing of some major conceptual convictions of the conservation movement necessarily is oversimplified. Certainly such concepts were not clearly and explicitly stated in all cases, nor did every member of the movement hold each of the concepts and ideas noted in our brief list. Furthermore, the various ideas and positions were by no means logically interrelated and mutually consistent; the amorphous and inclusive character of the 1890–1920 conservation movement precluded this. Nevertheless, we believe the foregoing paragraphs capture the essential content and spirit of the conservation movement.

The objectives of the movement flowed rather naturally from its conception. That first American conservation movement was a major attempt to change the broad goals, ideals, and purposes of the nation, so far as these affected the use of natural resources. The earlier ideology had been directed predominantly at private profit-seeking, at acquiring and exploiting natural resources for income or material benefit.

The conservationists rejected this as a sufficient or dominant objective. Instead, they sought to instill ideas based on concern for the future instead of wholly for the present, on stewardship instead of exploitation, and on some sort of higher duty instead of profit. They did not neglect either the technological implications or the physical attributes of resources, but they were usually less concerned with the economics of natural resource use, or they had economic ideas which today seem naive or quaint. In criticizing this deficiency, however, it is easy to forget that economics as a professional field of study and as an influence in public action was very much in its infancy before and during most of the years of the first conservation movement.

The methods used by the leaders of the conservation movement were diverse. First of all was the great use of propaganda or education—depending upon taste as to which term is more appropriate. Pinchot used this approach before he was involved in actual resource administration, and he always put it high on his list of priority measures; President Theodore Roosevelt was equally convinced of the efficacy of this approach, and was very skillful with it; indeed, many lesser figures in the movement were equally adept with it. The public had to be aroused, for only thus could public—political—action have sufficient support to prevail against the entrenched "interests." Moreover, only as people saw better lines of action could they be dissuaded from their old errors in their private use of resources.

In addition, public action of many other kinds ranked high in the strategy priorities of the movement, and President Roosevelt also was skillful on this front. Roosevelt firmly believed it was his duty to take any public action he thought essential or helpful to the general welfare, as long as it was not expressly prohibited by law. His extensive reservation of certain federal lands for use as power sites was not explicitly authorized by the existing law; neither was it forbidden. Believing strongly that the general welfare required the retention in public ownership of certain hydropower sites as well as other lands, Roosevelt withdrew them from private entry. President Taft, his successor, in addition to being much less convinced of the necessity of such actions, believed that he should do only what the President had been clearly authorized by law to do. The resulting personal and political conflict symbolized the difference between the ardent and the not-so-ardent conservationists.

Among the public actions taken during the general period covered by the conservation movement was the passage of the Forest Reserves Act in 1891. John Ise has shown how this was the product of a few intellectuals of the day, with neither popular support nor under-standing at the time.[5] Under the law, almost the full scope of the present national forest system was withdrawn from the public domain. President Roosevelt was responsible for larger withdrawals than any other President, including a large one made after legislation was passed in 1907, but not yet signed into law which prohibited him from making any further withdrawals in most western states. (Rumor has it, the withdrawal was made on the same day he signed the bill.) The system of federal wildlife refuges was begun during this same general period, important extensions were made to the national parks, and the national park service and the concept of a system of national parks was established in 1916. Other important legislation dealt with inland waterways, development and ownership of power sites, leasing of minerals from the public domain, and others. The story of this period has been told well and need not be repeated here.[6]

There were several reasons for the emphasis upon the federal lands. In the first place, such lands were larger in area, included far better land, and were relatively more important in the economy of that era than is true of federal lands today. A major congressional concern during the whole of the nineteenth century was public lands. In the second place, policies for the management and disposition of public lands were clearly proper fields for public action. Such policies had been formulated by the President and the Congress from the earliest days of the nation, and the public expected that policy would be made in this field. No President or Congress could avoid responsibility for action relating to the public lands. Third, there was serious concern, at least in intellectual circles, that both the past and current methods of land disposal were having undesirable and unanticipated results. The widespread trespass on public lands for harvesting timber and grazing; the unproductive nature of much land, especially forest land after both legal and illegal harvests; and the extensive tenancy on the farm lands of the time—all combined to throw into question the wisdom of the land disposal programs of the preceding decades. Fourth, there was considerable question as to the constitutionality of public programs for private lands, other than programs involving land grants and spending for education, and there was also great doubt as to the popular support for such programs. Even the most ardent

[5] Ise, *The United States Forest Policy* (New Haven: Yale University Press, 1920).
[6] Samuel T. Dana, *Forest and Range Policy—Its Development in the United States* (New York: McGraw-Hill Book Co., 1956). See it also for extensive further references.

conservationists doubted the legality of federal laws to control the use of private lands, and public ownership was not considered to be the answer to the serious problem of ailing agricultural soils.

The problems of private lands were not unnoticed by people in government at that time, however. The Country Life Commission, which met in 1908, found that farm life was harmed by various factors, among which was widespread soil erosion. The Conference of Governors held by the White House in that same year called attention to the need for prevention of soil erosion, but largely in relation to forests and the erosion on watersheds which would imperil the source of water for irrigation and would also cause siltation of reservoirs and irrigation canals.[7] An Assistant Secretary of Agriculture in 1894 had declared, "Thousands of acres of land in this country are abandoned every year because the surface has been washed and gullied beyond the possibility of profitable cultivation."[8] William J. Spillman, writing in 1910, insisted, "It is necessary that intelligent and vigorous effort be made to farm correctly. We must cease abusing the soil." And C. P. Hartley wrote in the same year, "More land has been rendered unfit for corn growing by the washing away of the surface soil than by constant cropping."[9]

Pinchot takes credit, probably accurately, for inspiring the famous Governors' Conference of 1908. At this massive Washington conference, every governor was present or represented, and scores of representatives from private organizations, members of the Congress and the Supreme Court, and just outstanding citizens attended. Following the Governors' Conference, a National Conservation Commission was established, and a North American Conservation Conference was held, at which Canada and Mexico were represented. Pinchot and Roosevelt had planned a world conservation conference, and invitations had gone out for a meeting at The Hague, but Taft killed the proposal upon his accession to the presidency. The Governors' Conference and its successors may well have produced some new insights into conservation. But, whatever else these various conferences did or failed to do, they most certainly brought conservation prominently to the attention of the American people, and to a lesser extent of the whole world.

[7] Pinchot, *op. cit.*, pp. 342 and 351.

[8] Charles W. Dabney in foreword to *Washed Soils: How to Prevent and Reclaim Them*, Farmers' Bulletin 20, U.S. Department of Agriculture, 1894.

[9] Spillman in *Soil Conservation*, Farmers' Bulletin 406, and Hartley in *Corn Cultivation*, Farmers' Bulletin 414, both U.S. Department of Agriculture, 1910.

Emphasis on the educational or propaganda aspects of the conservation movement, should not cause an underestimate of its solid achievements in terms of legislation or of the initiation of actual resource management. The beginnings of most present public resource programs can be traced to this period; the importance of this general era for resource programs can hardly be overestimated—even if it were not quite as great as Pinchot would have us believe.

Pinchot, Theodore Roosevelt, and the lesser conservationists aroused major conflicts over their programs. It could hardly have been otherwise. Not only were their programs new, and thus open to all the antagonisms that new programs or viewpoints naturally encounter, but also they conflicted with entrenched interests at many points. The withdrawal of the power sites, the establishment of the national forests, the trust-busting actions, and, above all, talk about the shortcomings of private enterprise in the handling of natural resources stepped on many toes and aroused many fears. The movement's leaders expected this reaction; they gloried in the opposition. To them, their programs were a crusade against evil, for the good and true, to be fought with every resource and means they could contrive. During the early part of this general conservation period, before Pinchot and Roosevelt had come to dominate the scene, there were justifiable complaints that the new national forests were "locking up" natural resources, for there was no way of selling the timber or other products to private buyers. While this legal deficiency was cured, those who wanted more favorable terms continued to criticize the forest program, particularly the way in which public lands and resources were later administered.

The controversies over conservation of resources reached their climax in a public quarrel between Pinchot and President Taft. Pinchot had been extremely influential with Theodore Roosevelt— perhaps less so than he thought, but certainly highly influential by any standards. Though Taft was Roosevelt's choice as a successor, he held considerably different viewpoints on conservation, as well as upon the role of the presidency generally. Between pique at no longer being in the inner circle and real concern over specific resource actions, or lack of action, Pinchot by his own account publicly sought a quarrel with President Taft. Naturally, he found it, and was dismissed by Taft from his position as head of the Forest Service. And Pinchot did not stop there; he went on to provoke a national controversy and a congressional inquiry into the administration of the public lands. This bitter controversy left scars that have hardly healed yet and have greatly affected federal land administration in the ensuing decades.

That story, however, is not central to our concern and we shall not pursue it further.

Even in retrospect, it is hard to evaluate the accomplishments of the conservation movement. By almost any historical test, it was a major social and political movement; it awakened a large public to resources generally, and to the need for conserving them particularly. As such, the movement was basic to much that has happened since the date we use as its close, including the subsequent development of the soil conservation programs which are the central focus of this book. Possibly the same results would have been accomplished in some other way, but the value of publicists such as Pinchot and Roosevelt should not be underestimated.

Neglect of Soil Resources

Although the conservation movement of 1890 to 1920 had many specific tangible public programs to its credit, it largely, and rather strangely, neglected soil conservation. This neglect seems odd because from the earliest colonial times, there had been a few men who were much concerned with soil conservation.[10] Such men were not only sincerely concerned, but did much through observation and experimentation to learn more about how to control erosion and other adverse changes in soils; also they often did a great deal to publicize their findings for the benefit of other farmers. These men were all private individuals, operating in a field which later came to be one of public action. Their contribution, as well as their devotion, was great, and the country owes them a great deal. However, in part, because most were ahead of their times; in part, because their pioneering work has been largely replaced by much more effective understanding; but mainly because as individuals there was no provision for continued effort—for all these reasons, their work did not evolve into a continued and effective program for soil conservation.

The conservation movement was not entirely silent on soil conservation, but it was nearly so. We have noted some instances of concern over it. While Pinchot occasionally mentioned soil erosion as a hazard, and soil conservation as a needed program his primary emphasis was elsewhere. It was natural that he should be especially concerned with forests and their use since his early interests had led him into forestry

[10] Angus McDonald, *Early American Soil Conservationists,* Miscellaneous Publication 449, Soil Conservation Service, U.S. Department of Agriculture, October 1941.

and this was where he obtained training and experience. But he gave more attention to the use of fuels, minerals, and waterpower than he did to soil conservation, although his training was no greater in those subjects than in soils. He simply did not give soil conservation high priority.

Marsh had paid somewhat more attention to soils in his 1865 book,[11] than did Pinchot at a much later period. With his overriding interest in ecology, Marsh recognized that soil was an integral part of the total ecological structure—what we today would call the "ecosystem," but he did not use that term. He recognized, and deplored, the effect upon the soil of deforestation, too much water resulting from accelerated floods, or unwise drainage. Marsh was not an advocate of any program, in the sense that Pinchot was to be a generation later, so we cannot charge him with neglecting soils. But he did not place the need for, or possibilities of, soil conservation programs high upon his list of resource issues.

The attitude of the leaders of the conservation movement toward soil conservation probably was affected by the common doubt in that period as to the legality of federal programs on private land. The constitutional issue had been raised in a major way in the consideration of the Weeks Act, passed in 1911, which provided for federal purchase of forest lands. The Act's legal underpinning was the power to regulate commerce given in the Constitution. It was then widely argued that forests exerted a beneficial effect upon stream flow, and thus aided navigation. Most of the public programs advocated and pushed by the conservation movement were on publicly owned lands or waters, where the legality issue did not arise. There were serious doubts about how any federal programs could be devised for private lands. On the other hand, there is little evidence that soil conservation programs on private land were seriously considered and set aside because of doubts over constitutionality.

The conservation movement from 1890 to 1920 was not alone in its relative neglect of the soil. Soil conservation also was largely, although not wholly, left out of the programs of the U.S. Department of Agriculture and those of the land-grant agricultural colleges, which were growing rapidly during this same period. Research and teaching had been authorized before the conservation movement of 1890–1920, but were greatly extended during this period; the agricultural extension system, in which the U.S. Department of Agriculture and the

[11] *Op. cit.*

land-grant colleges co-operate, began and was established on a national scale in 1914. The research-teaching-extension complex added greatly to the farmer's knowledge of production methodology; the housewives' knowledge of homemaking; and the nation's knowledge of the economic problems of agricultural production, marketing, price, and income. These were notable achievements. The seed for the agricultural revolution of modern times was planted during this period. But there was much less attention to soil conservation in either research, teaching, or extension; farmers and specialists alike, with few exceptions, were wrapped up in other problems. Soil conservation awaited a new prophet.

Chapter 3

THE SOIL CONSERVATION
MOVEMENT SINCE 1933

THE "great man" theory or approach to history need not be embraced
completely for agreement that strong individuals have helped mold
the course of events—in part by attracting others to them and their
ideas—and that individuals at times have symbolized political and
socioeconomic movements. We have discussed how Pinchot dominated
forestry and conservation generally from 1900 or earlier until perhaps
as late as 1920. We now come to soil conservation as such, where Hugh
Hammond Bennett played a dominant role for many years, from late
in 1920 until about 1950. For soil conservation, Bennett's role was as
dramatic and dominant as Pinchot's had been for forestry a generation
before.

Since every historical movement or change has its antecedents, one
is often at a loss exactly when to begin. But, as an effective public
program, soil conservation began with the New Deal, in early 1933.[1]
There had been some work on soil erosion and associated problems in
the U.S. Department of Agriculture and in the agricultural colleges
prior to 1930. Many men had written about the problem. Some had
developed practical means of constructing terraces, but this and other
work had never been organized around a central theme. It had little
continuity because it had been largely a personal effort of a few
individuals. In any case, it had been on a modest scale.

Bennett, long a soil surveyor in the U.S. Department of Agriculture,
became alarmed over erosion and resultant soil losses, particularly in
the Southeast with which he was especially familiar. He began preach-

---

[1] This chapter draws heavily upon Robert J. Morgan, *Governing Soil Conservation: Thirty Years of the New Decentralization* (Baltimore: The Johns Hopkins Press
for Resources for the Future, Inc., in press), and upon the references cited therein.
Our intent is to summarize rather than to add to Morgan's account.

ing and talking about the menace of soil erosion at various meetings and places in the later part of the twenties. By some adroit political maneuvering, which was characteristic of his later career, he managed to get the Appropriations Committee of the U.S. House of Representatives to push for a program of research on soil conservation, and to make a special appropriation for this purpose for fiscal 1930. Its chairman went directly to Bennett for ideas and testimony and so formulated the program that Bennett was the logical person to direct it.

At this time, as Bennett strongly pointed out, there was little clear and acceptable evidence on the nature and extent of the soil erosion problem. Moreover, as he stressed in testimony, there was a massive indifference to soil erosion throughout the country, by farmers and agricultural specialists alike. There were no generally accepted methods of dealing with erosion, even within relatively localized areas, much less on a national or a large regional scale. Various ideas had been advanced, some of which were strongly held by their advocates, yet there was far from complete agreement; some efforts had failed, tending to discredit the whole idea. For example, there was a notable lack of information on how to design terraces under different conditions.

The Coming of the New Deal

By 1933, the nation was in the throes of the Great Depression; the nation had been greatly weakened by this economic phenomenon. The number of unemployed workers ran from 8 million to 10 million; a comparable figure for today's larger labor force would be 15 million to 20 million. In agricultural areas, there was extensive tax delinquency on farm land; farm foreclosures were common; and there were cases of mob action to prevent foreclosures which would dispossess farm families. The country was upset and disturbed to a degree which it is almost impossible to believe today—difficult even for one who lived through the period, and still more so for younger people who were not alive or were too young to appreciate the circumstances. Many proposals for drastic economic and social change were advanced, and some were seriously considered by persons who would not, in other times, have listened to them. The nation was in a mood for change.

In early 1933, a number of proposals were made for major public works programs of soil conservation on private land as a means of employing men, especially unemployed men, in jobs which would protect and improve the land. There was an active bureaucratic war within the U.S. Department of Agriculture, and between that Department and the U.S. Department of the Interior, as to which agency and which person should carry out or direct these programs. We shall not trace the details of this struggle; it is sufficient to note that the direction wound up in Bennett's hands, in a new agency, the Soil Erosion Service, located in the Department of the Interior. During 1933 and the next year or two, an active program of public works on privately owned farm lands was pushed forward by this agency.

Also during the years 1933 to 1935, Bennett proposed and actively pushed for a major long-term soil conservation program, with broad powers to deal with the problem concentrated in a single agency. His proposals touched off intensive struggles within Agriculture, and between it and Interior; but one should bear in mind that such struggles probably would have occurred even if Bennett had not existed. The struggles had elements of personal power rivalries, which involved other powerful figures in addition to Bennett. Like most forceful leaders, he was ambitious; numerous persons then felt, and have since, that many of Bennett's maneuvers were dictated largely by a desire to enhance his personal position. However, there has never been any serious denial of his devotion to the cause of soil conservation.

In 1935, Congress passed legislation providing for soil conservation work on a permanent basis, establishing the Soil Conservation Service (SCS) as the agency to carry out the program, and transferring it from the Department of the Interior to the Department of Agriculture. Such legislation obviously had the approval of President Franklin D. Roosevelt, for at that time Roosevelt was so powerful politically that no legislation of this kind could have been passed without his concurrence and, in fact, his promotion. The process left Bennett more strongly entrenched than before, and he remained in charge of the program after its transfer to Agriculture.

The Act of 1935 was of prime importance in that it established soil conservation as an accepted national program—conservation of land for its own value, not merely as a means of employing idle men. Moreover, it was equally significant in establishing soil conservation as a special program, not as part of other agricultural programs. This latter decision is the more noteworthy because strenuous efforts were made by some agricultural interests to have soil conservation divided among numerous existing research and educational agencies. In still another respect, this legislation was far-reaching, for Bennett had made it clear in testimony that his idea of soil conservation was much more inclusive than merely building terraces to prevent soil erosion.

This 1935 legislation and the administrative organization embodied in it by no means ended the bureaucratic struggles over soil conservation; in fact, one may almost say it began then on a continuing basis. Under any circumstances, it would have been difficult to integrate into the U.S. Department of Agriculture, as then organized, an agency with broad powers to deal with a single major problem. But creation of the SCS represented a new approach, one difficult at best to synthesize into an established order of a different kind.

The bureaucratic power struggles which involved the SCS at that time were far broader than the single agency, or indeed than the Department of Agriculture. The Department had been basically a research and educational organization up to the time of the New Deal; it had worked primarily with agricultural colleges and, through them, with farmers. This resulted in a degree of local flexibility and adaptability which had many virtues. But the need to work through forty-eight state organizations made it impossible to carry out any national program on a uniform basis. In 1933, the Department of Agriculture began to deal directly with farmers under the Agricultural Adjustment Act, and later for the programs of the Farm Security Administration and for the soil conservation program. The result was a large measure of administrative confusion within the Department, struggles among different groups, and complex relationships to numerous outside organizations, including the land-grant colleges. Various efforts were made within the Department of Agriculture during the period to establish more orderly procedures and methods of operation.

A similar change in the relationship between the federal government and either local government or citizens was taking place with respect to direct public assistance grants, public work relief, and various forms of aid to cities, such as housing programs. In part, these changes in governmental relationships grew out of different concepts as to the proper role of the federal government, which President Roosevelt introduced, and in part out of the reluctance of state governments to assume a larger responsibility for dealing with the urgent social and economic problems of the day.

There were still other complicating factors. Many people in the Department of Agriculture considered soil conservation a much less serious problem than did Bennett and favored a much more gradual approach, primarily through education. All this contention was exacerbated by personal rivalries, which were perhaps inevitable; many resented Bennett's efforts, probably the more so because his efforts had been so successful. The other bureaus and agencies of the Department, and Secretary Henry A. Wallace made strenuous efforts to curb and delimit the scope of SCS activities. The co-operative Extension Service, within the Department but particularly in the colleges, made special efforts to prevent SCS from working directly with farmers—or, at least, from carrying out educational work with farmers other than those for whom it was doing special demonstration work.

So far as the general public was concerned, in the years immediately after 1933, soil conservation as an idea became enormously popular—even though many persons might have been hard put to define what they were in favor of. The massive dust storms of 1933, 1934, 1935, and 1936 had impressed and shocked the nation. Great dense clouds of dust rose from previously plowed fields, extended for thousands of feet into the air, and for hundreds of miles on the ground, often so thick as to make breathing itself difficult. The load of dust in the air could be seen as far away as the major cities of the Atlantic seaboard, and even out to sea for some hundreds of miles. Anyone, even the most obtuse, could hardly avoid seeing that something dramatic was taking place. The universal appeal of, and support for, soil conservation in those years was very great, perhaps never to be equaled later, although much greater understanding subsequently arose.

This situation had dramatic confirmation in early 1936. The Supreme Court declared the original agricultural adjustment program unconstitutional, and new substitute legislation was called the Soil Conservation and Domestic Allotment Act. It provided a system of grants in aid, or subsidies, to farmers who undertook specified farm programs called "soil conservation," although their production control aspects and aims were fairly obvious.[2] Thus was created a dual federal approach to "conservation," which has persisted with marked effects to the present. However, in early 1936, the important fact was that soil conservation had become a popular platform on which to base programs which under other circumstances would have been handled differently. In later years, the Agricultural Conservation Program (ACP) has had strong and effective leadership, but its activity cannot be identified with the personality of any particular man.

So far as we know, Bennett was unaware of our definition of natural resources as characteristics of nature that man knows how to use for desired ends. Had he known of it, he might have rejected it, for he did not have much patience with social scientists and social sciences. But,

[2] See Murray R. Benedict, *Farm Policies of the United States, 1790–1950* (New York: The Twentieth Century Fund, 1950), Chapter 14.

in effect, he was operating to change the natural resource formula considerably. He was acutely aware of the changes taking place in various aspects of nature, especially those arising out of soil erosion; he helped develop and spread new technologies for maintenance and improvement of soil productivity; he often asserted, usually without proof, that the measures he advocated were economically sound; and most of all, he sought to change the goals or objectives of land use—from operation solely or primarily for profit to stewardship or custo-dianship of an inherited soil resource which belonged basically to the whole people and which the individual only had a right to use.

Early Soil Conservation Programs

At first, the federal programs for soil conservation primarily involved demonstration projects on individually owned private farms or ranches. A farmer would enter into an agreement with the Soil Erosion Service—later the Soil Conservation Service—under which he promised to carry out certain measures and to permit the federal agency to carry out others. A major means of undertaking the federal activity was use of the manpower available through the Civilian Conservation Corps (CCC) or through the Works Progress Administration (WPA). In fact, one major purpose of the program was to provide such employment. Camps were established in local areas for the CCC, and sometimes for the WPA, and under the WPA program men were transported from small towns to work sites. Very large appropriations—for that day—had been made for various forms of direct public relief, and these moneys financed the soil conservation activities.

This type of effort was expensive in terms of the soil conservation results obtained per dollar of expenditure. Programs were often under-taken before there was a clear idea of what to do, or where, or how. The workers were sometimes inexperienced and were paid relatively little, even by current standards, and there was often a shortage of major equipment. The important consideration, however, was that the men be employed and paid, whether their use was timely or not. It was not difficult then or later to be critical of the soil conservation results from these programs, yet a good deal was accomplished—and their main purpose, in any event, was something else. Perhaps more serious, from the present point of view, is that such programs reached very few farmers directly. The indirect or demonstration effects were larger but cannot be clearly established.

Considerations of this kind led several men in the Department of Agriculture, including Bennett, Assistant Secretary M. L. Wilson, and others to look for a different approach. Wilson, at least, was familiar with the irrigation districts of the West, by means of which local groups of farmers undertook major responsibilities for land development programs which individually would have been beyond them. These districts provided a major vehicle with which the federal agencies could work. In addition to the irrigation and drainage districts, another example existed in the agricultural Extension Service, which a generation earlier had helped organize county farm bureaus in many states in order to have a farmer's organization with which to work.

As a result of extended discussions and research in the Department, a "standard act" for soil conservation districts was finally approved. It was published by the Department in May 1936, and shortly thereafter was strongly pushed by the Department, the President, and some agricultural interests. The purpose of the districts was to provide a means of reaching farmers that would be more efficient than individual contact, and to provide farmers with a mechanism for working together on problems of common concern.

The soil conservation district is a unit of local goverment, created under applicable state law.[3] In some respects it resembles a county, but it is more like an irrigation or drainage district in that it has special purposes or programs on defined areas only. The powers, functions, and organization of soil conservation districts vary somewhat from state to state, since they are determined by the laws of each. However, all the state laws were modeled after the standard act, modified as seemed desirable by the respective state legislatures.

Typically, the standard act authorized districts to carry on research for erosion control; to conduct demonstration projects; to carry out actual land conservation measures and programs; to enter into contracts with farmers and to give them financial and other assistance for conservation programs; to make gifts and loans of seeds, planting materials, and other supplies, and to lend equipment to farmers, for these purposes; to construct and maintain structures; to develop land-use plans for the districts and for farms; and to carry out other related programs. With some variations, most of these powers have been

[3] In 1937, state legislatures began adopting the standard soil conservation act, modifying it considerably in some cases, and by mid-1947 a district law was effective in every state. In Chapter 8, we consider in more detail the adoption of this standard act and the organization of districts under it.

extended to districts by the states. Although the standard act also provided for land-use regulations, this power generally was not
extended to the districts, and was not used extensively even where
granted.

Since the districts had no independent sources of revenue in most
states, they have depended mainly upon other units of government for
funds to carry out their programs. While, in general, they have been
authorized to accept help from any source, in practice, most of their
financial aid, manpower, and other help has come from the Soil
Conservation Service. Relatively small sums have been provided by
state and county governments in some states, and help has also been
extended by some other federal agencies.

Theoretically, local landowners and occupiers were to take the
initiative in forming soil conservation districts, which would become
formally established only after special elections. Formation of districts,
and to some extent operation of districts, had to be approved by a
state soil conservation committee, which was also supposed to help
promote the formation of districts. Actually, the formation of districts
has been promoted and aided by the Soil Conservation Service, and in
some states the Extension Service has aided in general educational
programs leading up to the formation of districts.

While the standard act was being drafted and adopted or modified
by the states, several policy issues arose. For example, how should the
state soil conservation committee be organized, and who should be its
members? According to the standard act, members would be the state
director of agricultural extension, the director of the agricultural
experiment station of the agricultural college, and a representative of
the Soil Conservation Service. In some states, perhaps because of fears
that this make-up would stop more action than it promoted, the membership of the state committee was changed, enlarged, or broadened,
particularly to include some farmers as members. Another policy issue
arose as to the power of the state committee. Could it refuse to
recognize a district voted in by landowners? How much power should
it, or does it, have to supervise or control the actions of districts? In
fact, the state committees have not been very effective—thus thwarting
the fears of their opponents and dashing the hopes of those who had
felt they would be a barrier to too much action. Precisely how shall
districts be established? Can tenants as well as landowners vote? What
constitutes a majority vote for approval of a district?

Still another range of policy issues concerned the legal powers that
the districts should possess. One feature of the standard act, adopted

by the U.S. Department of Agriculture after much discussion and weighing of pros and cons, was the power to enact land-use regulations. It was felt that situations would arise in which one or a very few farmers could refuse to co-operate in a district program and thereby jeopardize the efforts of all the other farmers in the watershed or area. Accordingly, it was felt that this regulatory power was needed, but should be used only as a last resort. Many states objected to it, and many state acts give no such power to districts. Where it was included in state acts, it has proved almost totally ineffective.[4] Farmers are most reluctant to impose land-use regulations on themselves or others, and under pressure have backed down even where regulations have been imposed. With few exceptions, districts have neither taxing power nor power of eminent domain. Although these powers were considered carefully by the Department, the standard act omits them on the theory that it would be hard enough to get enabling legislation passed without them, and impossible with such powers included.

Another issue in formulation of the act and its application was the size and boundaries of areas to be included within a district. In particular, debate ranged over whether districts should follow such political boundaries as county lines, or whether they should be based upon such natural areas as watersheds. Although the standard act was silent on the matter, apparently there was much feeling in the Department of Agriculture that natural boundary lines were preferable. In fact, however, the overwhelming majority of districts are organized on county boundaries; a few include more than one county, a few include only part of a county, but the county boundary is likely to be followed in either case, at least in part.

Soil Conservation and Other Programs since 1937

A number of developmental and organizational changes have affected the soil conservation program since about 1937. In that year, twenty-two states passed soil conservation districts acts with varying degrees of modification of the standard act. Districts began to be formed under the various acts almost immediately. The districts were organized in different ways, were staffed to the extent that the Soil Conservation Service could provide professional staff, and various programs have been carried out. Since then, in almost every year, laws

---

[4] The effectiveness of regulations in some Colorado districts led to their undoing.

were passed in additional states, more districts were organized in all states, more staff was added as appropriations permitted, and expanded programs were carried out within the districts. This process carried on through the war and postwar periods, with appropriately changing emphasis as external conditions changed.

Throughout this entire period, the SCS, the federal Extension Service, and most state extension services were engaged in a continued struggle—sometimes vigorous and sometimes more nearly quiescent, but never absent. A number of issues underlay this struggle, and personality clashes were not absent. One issue was over the formation of districts: Who should do the educational and promotional work with farmers? If the Extension Service, then soil conservation might well be submerged or shunted aside by a hostile extension agent or one too burdened with other programs. If SCS had this role, then a federal agency would be doing work that the state extension services had long zealously guarded as their own. Another question was: After a district was formed, what should be the relative role of each agency's technical personnel? Should SCS confine its efforts to working with farmers who asked for farm conservation plans? If so, might it be seriously boxed in, in many areas, by indifferent or hostile extension agents? But if SCS worked with all farmers in a district, what happened to Extension's traditional role of technical help to farmers? What if specialists from the two organizations disagreed on technical recommendations? And disagree they did, in numerous instances.

The conflict between these two agencies ebbed and flowed over this relatively long period. Some secretaries of agriculture—perhaps Clinton P. Anderson and Charles F. Brannan—favored SCS more than did others. Secretary Ezra Taft Benson, and through him, President Eisenhower, accepted advice from the Extension Service and from the American Farm Bureau Federation, and clipped the wings of SCS somewhat—perhaps as much as they thought they could without too violent a revolt in the Congress and in the country. Each agency or group obviously has had its supporters across the nation—otherwise, the battle could not have continued.

Also, SCS has had a long and continuing struggle within the Department of Agriculture, with the Agricultural Conservation Program (ACP), which has been administered by the Agricultural Adjustment Administration (AAA), the Production and Marketing Administration (PMA), and now is under the Agricultural Stabilization and Conservation Service (ASCS). As we have noted, the soil conservation program begun in 1936 actually was a means of making payments directly to farmers as income supplements, but in the name of soil conservation and sometimes in return for genuine soil conservation achievements. This raised basic controversial questions: Should payments be made for specific practices only, or for practices which fit into a complete soil conservation plan? If specific practices, what practices or requirements should be established for earning such payments? Who should define the acceptability of a practice, and who should check upon performance?

The ACP and crop control programs were organized to be administered by county committees elected by farmers. The Soil Conservation and Domestic Allotment Act, which established the program, originally made provision for the activity to be administered eventually by the states, but this provision has been repealed. The ASCS and its predecessors have stressed action, feeling that time and money could not be spared for long and expensive surveys or studies or plans; and, in practice, these agencies have been highly political in the partisan sense. The personnel of SCS has been recruited primarily from the Civil Service lists of eligible workers—most of them agricultural college graduates—who are moved into districts where their help is needed; during a good part of its history, it has only had field offices in counties where there were soil conservation districts. The SCS has strongly emphasized the complete farm plan, convinced that in many instances a specific practice might do as much harm as good unless geared to a long-run plan; and it has admittedly taken time to prepare such plans. At least nominally, SCS has been nonpolitical in the partisan sense, although at times it has been accused of political activity.

Every Secretary of Agriculture, from Henry A. Wallace onward, has been concerned over the divergence of these programs. The inherent conflict became an actual one, with increasing intensity up to the postwar years. Secretary Anderson and Secretary Brannan conducted long and unsuccessful efforts to harmonize the programs of these two agencies. Each agency felt itself too deeply committed to its programs to yield to the other, each had sufficient farmer and political strength to feel it did not have to surrender; and, between them, they thwarted efforts of the secretaries at co-ordination—not openly, but effectively. In 1947 and 1948, scandalous open warfare broke out between these two agencies in several states. Congress—more particularly, the House Appropriations Committee, and, even more precisely, certain influential congressmen who foresaw nothing but disaster for agricultural programs in general if such warfare continued—forced a compromise in 1951. With his hand thus strengthened by congressional action, Secretary Brannan was able to bring about a large measure of co-ordination between the two agencies. One major step was to provide for the transfer of limited funds from PMA to SCS, so that SCS could employ enough technical people to check the performance of farmers under PMA programs. On its part, SCS was required to provide some services to farmers who were not in soil conservation districts or were not co-operating in district activities—which it had been reluctant to do.

The administrative arrangements formalized in 1951 have continued. In addition to providing a modus operandi for co-operation, the congressional action made each agency realize that its dominance would not be tolerated—that each agency was needed and had a job to do, and that their conflicts had to be resolved.

One other internal matter in the Department of Agriculture was the relationship between the SCS activities, particularly the conservation planning and land classification activities, and the soil survey activities of the Department. The latter had been in the Department's Bureau of Chemistry and Soils when SCS was established. There had long been complaint on each side—the soil surveyors felt that SCS technicians were not adequately qualified and that much of their work lacked scientific validity, SCS felt that much of the soil survey work was too technical for use by farmers. In 1952, the soil survey activities were transferred to SCS, and internally it has clarified the relationship between these programs.

Over the years since 1937, some new programs have been added for SCS to administer. Among activities for which SCS has had major responsibility are the Small Watershed Program since 1954, and the Great Plains Conservation Program since 1956. While these, like any new public programs, involved some problems for the agencies concerned, these two did not touch off new major bureaucratic struggles within the agricultural field. (SCS had a brief but intense political struggle, however, with the Army Corps of Engineers as to who should be responsible for the Small Watershed Program.)

We, and most others reporting on the development of soil conservation programs in the United States, have given most attention to policy issues and political struggles over public programs. These events are obvious and easily documented. But there were other public actions, and numerous private ones, during these same years that also deserve some attention. One important event of this kind was the formation of the National Association of Soil Conservation Districts (NASCD) in 1946. By July 1, 1946, there were 1,638 soil conservation districts in the United States. From one point of view, districts in different parts of the country have similar problems and, hence, an obvious advantage in exchanging ideas and information. From another, all districts would like SCS to get adequate appropriations so that it may properly aid the districts. Any private pressure group directly concerned with a particular public program can be accused of merely serving the interests of the agency—and sometimes, or to some degree, this may be true. But such a group also can be accused of dominating the operations of its chief interest agency, until the latter is little more than its tool—and sometimes, or to some degree, this also may be true. Paradoxically, each may be true at the same time to some degree. That is, the pressure group serves the interest of the agency while at the same time serving its own interests, and the agency is certainly not oblivious to the attitudes of the pressure group. This kind of relationship has existed between NASCD and SCS; the former has supported appropriations for the latter, but the latter is surely not oblivious to the interests of the former.

The great influence of private activities, along with public programs, is especially marked in the field of education and information. Before 1933, there was relatively little information about soil conservation available to farmers or others interested in the subject, and almost nothing of a regular and continuing nature. Today, the situation is drastically different.

The National Association of Soil Conservation Districts, as an example, has published a weekly *President's Letter* (or *Tuesday Letter*) since 1952. This goes to 32,000 soil conservation district directors, to SCS and ACP personnel, to county agents in the Extension Service, and to many others, including members of Congress. It is concerned with internal organizational matters of the National Association and its state associations; with program developments in the general field; and, to a limited extent, with legislation and appropriation matters. Since it is frequent and regular, it serves almost as a newspaper in this field. Also, NASCD publishes a *Watershed Letter*—sent to about 12,000 names (one copy to each district, instead of one to each director for the *Tuesday Letter*)—which is concerned primarily with the Small Watershed Program. It also publishes a *Great Plains Newsletter*, which goes to about 7,000 names and is concerned with that special program. In addition, in a number of states the State Soil Conservation Committee, or the State Soil Conservation Association, or both, publish regular letters.

The Soil Conservation Society of America, formed in 1943, publishes the *Journal of Soil and Water Conservation*. A membership organization, it has chapters in all states and in other local areas, with a total of 114 chapters. Its membership of 11,000 is drawn primarily from the ranks of personnel employed by public agencies and engaged directly in soil conservation work. The Society is noteworthy for the breadth of its definition of conservation, including not only soil but all forms of vegetation and other beneficial plant and animal life. Its interests are primarily technical and professional, and its *Journal* has readable and interesting articles on various aspects of the art and science of soil conservation. In addition, the Society has a number of technical study committees. As a nonprofit organization, it does not undertake political actions, but the Society undoubtedly has been a major means of communication among those persons professionally engaged in soil conservation.

The SCS publishes a monthly magazine, *Soil Conservation*, largely for its own personnel and for co-operating agencies. Begun in 1935, it now reaches about 10,000 copies monthly. To receive copies, district directors must subscribe, and many do; free distribution within and outside SCS is limited to 7,500 copies. The monthly deals mainly with examples of soil conservation actions, methods, and programs, as applied on farms, but has also included some book reviews.

There has also been a relatively large flow of bulletins, pamphlets, and other publications relating to soil conservation from various parts of the Department of Agriculture and from the agricultural colleges. The typical college publication is issued as a one-time type of report not a regular periodical, and the distribution has varied according to subject and intended audience. In addition, a number of citizen organizations in the general field of conservation have shown increased interest in soil conservation over the past thirty years or so. These include the American Forestry Association, the Izaak Walton League, the Audubon Society of America, Nature Conservancy, and others. Also, several of the professional organizations have increasingly dealt with soil conservation in their annual meetings and professional journals. Manufacturers and distributors of farm machinery, construction machinery, and other related products have also taken increased interest in soil conservation, often conducting educational programs, offering prizes and scholarships, and in various ways helping to arouse the interest of farmers and the public. The soil conservation districts also serve as a channel of information to and among farmers. By meetings and through informal discussions, the ideas of conservation leaders have been disseminated among farmers, and ideas and viewpoints from farmers have been relayed back to conservation specialists. By rallying local leadership behind soil conservation, the district form of operation has helped to spread the acceptance of the public soil conservation programs.

Most of these information programs, however, reach those already converted or favorably disposed to soil conservation. The vast general public still is much less informed, and much less interested, although at intervals magazines of general circulation, radio, movies, and other mass media use soil conservation as a theme to which there is a relatively good public response. The channels of communication within the general soil conservation field, on the whole, are good. All of this is a far cry from 1933, when Bennett properly stressed ignorance and indifference about soil erosion as major problems.

Summary

In this brief historical review of federal soil conservation programs, many details have been omitted, as well as supporting evidence; and some events or episodes which seemed critical at the time have been omitted entirely because they seem less important in a brief retrospective examination. But a few major conclusions stand out:

1. From the start, SCS has had to fight—first, for its birth; continuously, for its existence. The intensity of the struggle has waxed and waned over the years. Very recently and right now, the front is peaceful—in the same sense that there are lulls in the Cold War. The SCS has had to build and maintain political and popular strength, and to use any means necessary to the attainment of the goals it considered critical; it has never been able to relax its guard, to consider methods of operation not built on strength.

2. The SCS has brought a dedication and a single-mindedness to the problems of soil conservation, without which the nation almost certainly would not have realized the significant soil conservation achievements which we shall consider in Chapter 8. It has gradually developed a more nearly integrated professional or technical approach to the problems of soil conservation. The contributions of engineering, hydrology, agronomy, soil science, and other fields of technical knowledge have been fruitfully brought to bear in a more nearly coordinated fashion.

3. Largely as a result of its need to fight for existence, the SCS has had to develop a program which promised something for everyone, or at least for every farmer. It needed political and popular support in

every state and in every congressional district, and it has actively sought such support. Some of its programs—its technical help on drainage problems, for example—seem more clearly designed to help win such political support than to contribute to soil conservation. The agency has never felt free to concentrate on the most serious erosion or other problem areas, but has had to spread its resources everywhere.

4. The SCS has never had the power and the authority to deal with the problems of agricultural adjustment, broadly defined, even when such adjustments might have profound consequences for soil conservation. In many situations, the best soil conservation program might be to take the land right out of agriculture, but SCS has had no authority to stimulate and aid such shifts. When it has suggested that a piece of land might better be shifted from crops to pasture, the question has arisen: what is the farmer to do with the grass? He may have to add livestock, change his whole way of farming, which, in turn, may call for training and credit. And SCS has had no authority to help with any of these adjustments. Had SCS undertaken any programs along these lines, the agency rivalries would surely have flared up anew; yet there has been ineffectual co-operation with other agencies to deal with such problems.

5. The SCS has had to work with farms and farmers as it found them. The need to develop soil conservation programs for a particular farm, as currently bounded, and for a farmer, perhaps with serious limitations of managerial ability and capital, has imposed severe technical and socioeconomic strains on planning. The result has been a limited "success," even by SCS standards, and still more limited success by national or social standards.

6. An inflexibility in national agricultural programs, one which is not easily modified, has grown up. The powers of SCS are important but limited; and SCS understandably is reluctant to yield them. Yet SCS cannot deal with many problems, which other agencies either cannot or do not deal with either. One need not criticize SCS or any other agency, and yet be seriously concerned to ask if this arrangement is in the national interest for the next generation or longer.

Chapter 4

EVOLVING CONTENT OF THE SOIL CONSERVATION PROGRAM, 1930-1964

THE underlying assumptions and philosophy of the soil conservation program in the 1930–64 period bear striking similarities to the first conservation movement of a generation earlier. Like its predecessor, this soil conservation movement has been neither sharply and clearly defined nor formally conceptualized. In spite of a great deal of attention, or possibly because of a spate of talk and writing of a rambling and discursive kind, the logical and factual base of soil conservation often has been unclear and confused. Soil conservation includes a very broad range of ideas, not always completely harmonious. Different people within the movement have held different ideas; the outsider has been forced to draw his own conclusions, which have not always been the same as those within the movement. At the least, one can say that soil conservation of this general period has not been strictly and coldly economic, either in its motivation or operation.

On the contrary, soil conservation was an emotional matter to those who were active in the early stages of the program. To those dedicated persons, the loss of good topsoils through erosion was emotionally disturbing. To them, it was a matter of ethics that the soil should be protected, and that it should be left in as good or better condition than when it was received. They emphasized stewardship, or care of soil resources entrusted only for use to a particular generation. Many ardent conservationists expressed genuine horror at soil erosion—often reflected by such expressions as "rape."

57

It is probably significant that Hugh Hammond Bennett, the apostle of soil conservation, was often asked to preach in rural churches, especially in his native Southeast. To these farm people, and to him, soil conservation was a religious matter, a matter of faith and ethics, not merely of economics, good business, and the comparison of costs and returns. They thought, as he did, that it was wholly natural to use houses of worship as forums for exposition of his convictions in this direction. He brought to soil conservation a religious conviction and fervor—and also, some would say, a religious intolerance.

Particularly at the beginning of the soil conservation program, as in the earlier general conservation era, there was a strong, though not explicitly stated, strain of Malthusianism. The point was made over and over that the nation will one day need all its productive land and topsoil; that it cannot afford to lose any soil which will never be replaced within modern lifetimes. This viewpoint was the more remarkable because at the same time the view was prevalent—mainly in other quarters—that the United States was rapidly approaching a stationary population. However, it is true that up to the nineteen thirties there had been little reason for expecting the phenomenal rise in crop yields and other production per acre which came later, and additional land was then viewed as the chief, or only, means of increasing agricultural output to meet growing needs.

The soil conservation movement, while similar to the earlier general conservation movement in many ways, did not include as wide a range of ideas and subjects. For example, it was free of worry about monopolies, immigration, and child labor laws.

Congressional Intent

The soil conservation movement considered here was broader than a federal program, yet the federal program was its major part. The federal program included more than legislation, yet the laws and the appropriations were significant, partly as an expression of prevailing attitudes and partly as a means of granting power to crusading federal agencies. Various acts were passed, and there was much legislation. A full review of congressional policies and attitudes would take us too far afield,[1] but we can briefly note a few.

[1] See Robert J. Morgan, *Governing Soil Conservation: Thirty Years of the New Decentralization* (Baltimore: The Johns Hopkins Press for Resources for the Future, Inc., in press). See also Hugh Hammond Bennett, *Soil Conservation* (New York: McGraw-Hill Book Co., 1939), especially pp. v–xi.

The earliest appropriations specifically for soil conservation were made in 1928 and became effective for the fiscal 1930 year. These were explicitly concerned with research on soil erosion and its prevention, and the major emphasis was upon use of terraces for its control. The New Deal's earliest soil conservation activity was heavily motivated by the need not only to provide public works for the large numbers of unemployed men, but also to deal with the severe dust storms and drought conditions—both of which were believed to be largely temporary. However, the preamble to the Soil Conservation Act of 1935 contains a major expression of national policy:

> It is hereby recognized that the wastage of soil and moisture resources on farm, grazing, and forest lands of the Nation, resulting from soil erosion, is a menace to the national welfare and it is hereby declared to be the policy of the Congress to provide permanently for the control and prevention of soil erosion and thereby to preserve the natural resources, control floods, prevent impairment of reservoirs, and maintain the navigability of rivers and harbors, protect public health, public lands, and relieve unemployment.

Morgan has emphasized the importance of this legislation.[2] First, it explicitly recognized erosion as a sufficiently serious national problem to warrant a permanent and nationwide program of action. Second, it expressly directed that the program be comprehensive. Third, it endorsed the idea of program planning for soil conservation. It was based upon the assumption that the erosion problem could be identified, isolated, and treated. But it also made clear that the task called for adjustment to a host of complicating circumstances, political and economic.

The nature of the program undertaken under this broad grant of authority has changed gradually, but greatly, over the intervening years, and additional legislation has been passed. Yet most of the changes may be considered as means for working out the broad principles and concepts included in the basic 1935 legislation. There has been no repudiation of the basic legislation and concepts, but rather a natural flowering of them.

Within a year after passage of the 1935 soil conservation legislation, the Soil Conservation and Domestic Allotment Act of 1936 was passed.

[2] *Op. cit.*

Its preamble spelled out five purposes:

(1) Preservation and improvement of soil fertility;
(2) promotion of the economic use and conservation of land;
(3) diminution of the exploitation and wasteful and unscientific use of national soil resources;
(4) the protection of rivers and harbors against the results of soil erosion in aid of maintaining the navigability of rivers and harbors against the results of soil erosion and in aid of flood control;
(5) re-establishment . . . of the ratio between the purchasing power of the net income per person on farms and that of the income per person not on farms that prevailed during the five-year period August 1909–July 1914, inclusive.

In some respects, this language was broader than the statement of objectives in the 1935 Soil Conservation Act (to which, it was technically, an amendment). There is more concern with the level of inputs into an existing production function—note the reference to soil fertility. There is also more recognition of building basic soil productive capacity, as well as its maintenance. It is not clear how far, if at all, Congress and the Administration intended to broaden the scope and content of the soil conservation part of this law, and how far they were merely reaffirming objectives stated differently in the original act.

The 1935 Act has been the legislative charter for the program of the Soil Conservation Service (SCS), while the 1936 Act has served similarly for the Agricultural Conservation Program (ACP), now administered by the Agricultural Stabilization and Conservation Service (ASCS), successor agency to the Agricultural Adjustment Administration (AAA) and to the Production and Marketing Administration (PMA). All of the programs are and have been in the U.S. Department of Agriculture.

The Early Soil Conservation Program

The first emphasis in the evolving soil conservation program was on research. This was almost the only approach in the days up to 1933, at least at the level of the federal agency. Bennett knew, from his extensive field experience as a soil surveyor, that the causes of soil erosion were numerous and their interaction complex. He also realized that the methods for control of erosion would have to be varied from place to place to meet the peculiar conditions of each. It was also clear that critics would be quick to seize upon mistakes, and that a sound research base was necessary to avoid mistakes and to convince the doubters. Moreover, in that period, research was eminently respectable in a professional and political sense, when more direct action would not have been. Accordingly, Bennett's first approaches were to enlist support for specific soil erosion research.

While any observer of farm land knew that some crops protected the soil against erosion better than others, this general fact needed spelling out for different soil types, slopes, crop rotations, cultural practices, and other circumstances. The first research, therefore, consisted largely of trying to measure soil losses under widely varying experimental conditions. But even to measure soil loss was not simple. The methods of trapping and measuring soil washed off experimental plots are fairly standard today, but prior to 1930 they still had to be perfected. In fact, the research on soil erosion and its control was barely started when the New Deal came along, with its immediate emphasis upon action programs of soil conservation. During the early years of these national programs, critics were constantly pointing out that action rested upon a very slender factual and research base.

With the coming of the more active phase of soil conservation in 1933, emphasis quickly shifted to surveys to measure the seriousness of soil erosion in different parts of the country. Again, anyone moderately familiar with agriculture in the United States knew that some sections of the country had a more serious erosion problem than did others; but measurement of the degree of seriousness of the problem and identification of the areas which had different kinds and degrees of erosion were large tasks. Even to define the kinds and amounts of erosion was difficult. One of the first steps of the newly created Soil Erosion Service had been to undertake a national erosion survey. In this, it was encouraged by the new National Resources Committee (later to become the National Resources Board and then the National Resources Planning Board), which was interested both in the problem and in the employment possibilities of public works for erosion control.

The first national estimates were published by the National Resources Board in 1934 in two forms—in a map (part of which we reproduce in Figure 6) and as statistics (which we present in Table 13 of Chapter 8). This Reconnaissance Erosion Survey had been done quickly and under pressure; the results could be but approximations, and there was a general belief at the time that Bennett and his

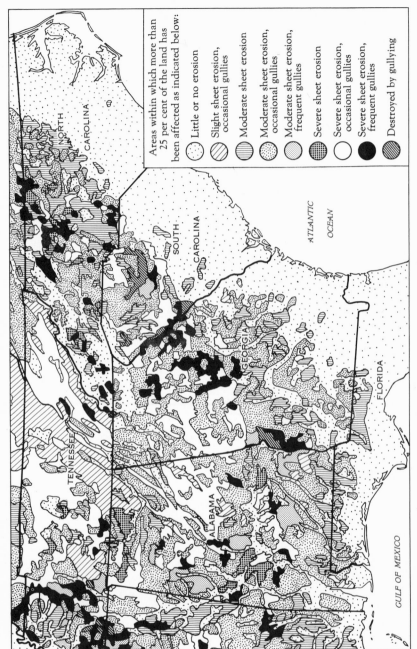

Figure 6. Erosion in the Southeast, Reconnaissance Erosion Survey, March 1935. (Taken from U.S. map in *Supplementary Report of the Land Planning Committee to the National Resources Board*, Part V, facing p. 22.)

colleagues seriously exaggerated the problems they encountered. Yet this first national survey was, and remains, a landmark, partly because it was the first comprehensive national report upon the general problem. Moreover, the results as revealed by that first map have stood up pretty well compared with later studies of the soil conservation problem. (See, for example, our Figures 23 and 24 and discussion of later studies in Chapter 8.) This first national survey had another virtue; it represented a degree of generalization which SCS was reluctant to make subsequently, as it learned more details of the problem and as its early work came under criticism for errors in specific localities. Furthermore, these general national estimates helped greatly in dramatizing the soil erosion problem as national concern with the issue mounted.

The emphasis in the early soil conservation programs was almost exclusively upon erosion, as a threat and as a problem to be dealt with. This is found in Bennett's early statements to congressional committees and the various documents he prepared for the Secretary of the Interior, Harold L. Ickes and others; it is found also in the Department of Agriculture's committee report in 1935 on how to fit the soil conservation program into the Department's administrative structure. In addition, it is clear in the reports of the Land Planning Committee of the National Resources Board; the annual reports of the Soil Conservation Service for 1935, 1936, and 1937; in *Soils and Men—The Yearbook of Agriculture, 1938;* and in Bennett's own 1939 book. All repeatedly emphasized soil erosion.[3] We stress this early preoccupation with the erosion of soil because the emphasis was later to change markedly.

In the early period of emphasis upon soil erosion, much was made of the extent of the soil losses—the areas destroyed for future cropping, the areas severely damaged, the tons or acres of topsoil going down the major rivers daily, and so on. Erosion was often equated with sin—something everyone was against with no questioning. As a result of this emphasis in many of these statements, qualifications to the effect that some areas were actually gaining soil materials by accretions, that some areas were untouched by erosion, and that others were experiencing rapid but geologic erosion unaffected by man often were

[3] See Morgan, *op. cit.*; and *Supplementary Report of the Land Planning Committee to the National Resources Board* (published in eleven separate parts), Part V, "Soil Erosion: A Critical Problem in American Agriculture," prepared by the Soil Conservation Service and Bureau of Agricultural Engineering of the U.S. Department of Agriculture for the Land Planning Committee (Washington: Government Printing Office, 1935); *Soils and Men—The Yearbook of Agriculture, 1938* (Washington: Government Printing Office); and Bennett, *op. cit.*

lost or overlooked by many readers. This was the crusading era—and no crusader can stop to weigh opposing evidence and arguments for fear of diluting or losing his crusade.

During this same period, and largely in these same publications, the primary emphasis is placed on terraces as a device for erosion control. Bennett and many of his close colleagues always recognized the need for a broader approach; but even they, when asked for specific examples by congressional committees or others, nearly always offered terraces as the chief illustration. Others, notably the Bureau of Agricultural Engineering of the Department of Agriculture, looked upon terracing as the chief, if not the only, approach.[4] Critics then and later pointed out that terracing was not a universal remedy—that there were land situations where terraces were impractical. Moreover, some terraces had been installed which did more harm than good, because they tended to collect and channel runoff water in ways that created more erosion than would have occurred without them. In modern soil conservation programs, terraces are one, but only one, of many important practices.

In that early period, also, the soil conservation program suffered from the view held by some that it was chiefly a short-term public works program for the employment of idle men. In fact, at one stage there was grave danger that the program would pass out of the hands of the soil conservationists, or more particularly of Bennett's hands, into the control of the Federal Emergency Relief Administration (FERA), which had been established in 1934 to help relieve hardships caused by unemployment and the drought. The public works aspect of soil conservation certainly was a critical part of the whole New Deal program of that period, and since the early soil conservation programs were often poorly developed or even nonexistent, much manpower and expenditure were wasted or had a poor output. Yet the record probably was no worse than for public works elsewhere at the same time. The nation had started its massive public works programs with little or no advance planning because the demands for employment were too urgent to wait for the necessary planning. Moreover, a great deal of good was accomplished, and the programs improved as time went on.

[4] See *Supplementary Report of the Land Planning Committee to the National Resources Board,* especially pp. 96–106.

The Addition of New Program Elements

After the Soil Conservation Act of 1935, a number of ideas, concepts, and components of the total program evolved more or less concurrently, and were mutually interacting. Although we necessarily must discuss them one at a time, they did not occur in isolation. In discussing the evolvement of the program during the second half of the thirties, our major emphasis will be upon the programs of the Soil Conservation Service, but the 1936 Act provided substantial ACP payments to farmers for carrying out conservation practices (which we discuss in detail in Chapter 8), and state and local governments increasingly came to have soil conservation programs, although on a comparatively small scale.

All of the programs, to be effective, had to be carried out on land owned or managed by someone, and of land affected only a relatively small area was publicly owned. During the thirties, the federal government purchased several million acres of land judged to be submarginal for continued agriculture, and on this land conservation programs were carried out through the various public works programs of the period. Most of the public soil conservation programs, however, had to be carried out on privately owned farm land. The co-operation of the farmers was thus necessary, and was eagerly sought. Farmers were given numerous incentives to encourage them to undertake conservation measures. Much of our discussion is in terms of public programs— partly because it is easier to get information about them than about the critical role of private landowners. As time went on, an unknown but presumably growing number of farmers undertook soil conservation programs independently of any public programs. However, since the amount and variety of public assistance for soil conservation was relatively great, probably most private landowners had some form of public help.

One change during the thirties was the improvement in making soil erosion surveys. There was better recognition of the amount and rate of soil erosion from the land; this, in turn, was related more closely to the natural conditions of the site and to the land-use practices. While this may seem a small gain, it led to more accurate surveys of soil erosion and to better professional and public acceptance of those made.

Concern over soil erosion surveys and the many different factors affecting soil erosion was largely responsible for the development of the system of land-use capability classification. Various facts about the slope of the land, soil type, past erosion, vegetative conditions, general susceptibility of the land to erosion hazards, and other factors were combined into a single rating. (The nature of this classification and the factors that enter into it are discussed in more detail in Chapter 6.) The precise definitions of the various factors varied from place to place, depending in part upon other factors, such as climate, not specifically included in the formula. This variation in definition created a good deal of confusion at first, and there were some controversies over the rating given to specific local factors. Others tried to employ this land-use capability classification as a measure of land productivity; in fact, while some of the factors are common to both classification systems, there is a good deal of difference between them. As land-use capability became better understood and perhaps more accurately measured, it became evident that the general, or summary, classification into eight broad land-use capability classes had great utility. The land-capability classification seems to have been mentioned first in the *Report of the Chief of the Soil Conservation Service, 1939,* the 1940 report of the SCS Chief has an extended discussion of the system then in use. Bennett, in his book published in 1939[5] and presumably written at least a year earlier, mentions such classifications only briefly.

During these same years, in addition to terracing, the first major practice, other measures to control soil erosion were developed and put into use. Emphasis was placed on plowing and cultivation on the contour, the use of different crop rotations, the use of stubble mulch in the drier regions, greater use of pastures, and various other agronomic practices. Emphasis upon the use of protective crops naturally led to the establishment of forest and grass nurseries, where suitable planting stock and seeds could be grown. Previously, many new species had been totally unavailable to farmers, or unavailable in adequate quantity, and seeds or vegetative planting materials of native species were unavailable in sufficient volume for farm use. Plant introductions were made in some cases, but more often plants which had been previously introduced were developed and distributed for farm use. Among the latter, for instance, was crested wheat grass which has been by far the most common type of grass sown on depleted western range lands. At the same time, methods were devised for the harvesting of seed from native grass and other plants. By the end of the thirties, SCS had an extensive system of nurseries, which

[5] *Op. cit.*

was reduced somewhat during the Eisenhower administration, and which now continues on a level adequate to provide seed supplies and planting materials.

The SCS has not served primarily as an agency to import new species from abroad or to breed them, and it has not acted as a volume seed producer. Rather, its role has to consider the needs for new plants for soil conservation purposes—grasses and shrubs primarily. Then it has sought to find plants adapted to the need, and to have them tested by farmers under field conditions. The SCS nurseries have made initial tests and produced foundation stock. Seeds for proved varieties are then produced by commercial producers, but SCS usually maintains sufficient foundation stock to insure the purity of the variety. The field testing by farmers is worked out co-operatively with the soil conservation districts concerned. This program has been quite effective, especially in the West, as a means of making available to farmers and ranchers plant varieties especially suited to conservation needs.

During these years, the idea of the soil conservation district was developed, the "standard act" was drafted by the U.S. Department of Agriculture, state enabling acts were passed, and districts were formed and started operations. Districts were to become a major mechanism for carrying out much of the general program of soil conservation, but they are a means rather than a program in themselves. During much of this time, and later, however, the promotion and servicing of new districts became a program in and of itself for the Soil Conservation Service, absorbing no small amount of manpower and resources. (More specific information on the activities of districts is presented in Chapter 8.)

During the thirties, also, the idea of the whole farm conservation plan was developed and became deeply embedded in the thinking and operations of Bennett and other SCS leaders. They became convinced that separate and unrelated soil management practices on a farm could not add up to a real program of soil conservation; in fact, that unrelated practices could sometimes do more harm than good. They were convinced that the way a farmer used one piece of land depended in part upon the other land in his farm—how it could be used, and how its use affected the tract in question. They felt that a conservation plan for a farm should grow out of, and be based upon, a classification of land-use capability for that farm. Their philosophy for land use and treatment was dramatically stated in the 1939 annual report of the SCS Chief, as follows:

The use of every acre of every field, pasture, and woodlot of every farm according to its individual adaptabilities; the treatment of every acre of every field, pasture, and woodlot of every farm according to its individual needs.[6]

The similarity, not to say paraphrase, of this statement and the creed of the Socialists is striking, but almost certainly unconscious. It can, however, well be described as a form of socialism for land.

Under this concept of whole farm planning, a land-capability classification was made; programs of land use were proposed for the different kinds of land and for the different fields. These programs included special soil conservation measures, such as terraces, grassed waterways, and others, and special crop rotations and practices. The plan naturally had to get into such matters as the use of pasture and feed crops for livestock, the use of crops on the farm or their sale, methods of farm management generally, and the like. If realistic, the plan had to consider not only the limitations of the land included within a farm, but also the ability of the farmer to carry out various practices, including his capital or credit. This approach was certainly adopted in the firm conviction that anything less was unsatisfactory and could be highly disappointing.

But this whole farm planning by SCS resulted in many controversies with farmers, PMA, and the Extension Offices. The causes of each of the controversies were different. Farmers were often interested in only a single practice. For example, a farmer might want technical help in laying out a system of terraces on his farm, but in nothing more. He might tolerate the whole plan in order to get what he wanted, but he often followed only that part in which he was most interested. He was inclined to think that the other parts of the plan were unnecessary, impractical, or beyond his ability. On its side, SCS was reluctant to have only part of an integrated plan followed for fear that it would prove partially or wholly unsatisfactory. These differences between SCS and the farmers generally did not flare up into sharp controversies.

But the same basic differences existed between SCS and PMA. The latter wanted "action," and wanted to make payments to encourage specific practices. The SCS process of farm planning took so much time, in relation to available manpower, that it could reach only relatively few farmers each year. The PMA personnel tended to think of SCS plans as theoretical and unworkable or unrealistic, while SCS

[6] *Report of the Chief of the Soil Conservation Service, 1939*, U.S. Department of Agriculture.

thought that much of the PMA program was wasted. This controversy lasted for a long period.

With the Extension Service, the basis of the controversy was different. If SCS carried its whole farm plan to its logical conclusion, it got into such matters as livestock numbers, feeding programs, farm incomes and expenses, and the like. In fact, only by considering such matters could the name "complete farm planning" be justified, or could the planner be sure that his proposed plan was the best one or at least a good one. But such planning got into the kind of farm management and other planning in which the Extension Service had been engaged for some years, at least in some states—and so the cry of duplication went up. This provided fresh fuel for the controversies with Extension that would almost surely have arisen in any case.

The foregoing discussion is largely in terms of cropland. However, during these same years, SCS, the ranchers, and others also began to make notable progress toward conservation of range lands. The same basic techniques were applied to range land, and SCS developed considerable skill in working with ranchers and in enlisting their cooperation in conservation programs. During this same period also, some progress was made toward better forestry practices; on farms, this was largely a result of SCS farm programs.

A Shift to Land Management, 1935 to 1950

Gradually during the general period 1935 to 1950, and to some extent subconsciously, the emphasis of the whole group of soil conservationists, in both public and private programs, shifted from the control of soil erosion to the management of the land for greater productivity. This was in many respects a natural evolution, yet it greatly changed the basic purpose of the soil programs, especially when viewed from a national or social point of view.

The nature of this change of emphasis may be illustrated this way: A farm is experiencing severe sheet and gully erosion, caused primarily by excessive runoff from steep fields cultivated to row crops. As one major means of coping with this problem, terraces are installed to slow down runoff and to lead the water off the land over grassed waterways. At the same time, contour cultivation and new crop rotations are introduced. Soil loss drops from a level which in time would have destroyed the land for cropping, to a level which is more or less balanced by the formation of new soil from parent soil material. Crop yields may drop temporarily, due to exposure of raw soil, but soon recover. So far, so good; this is clearly an erosion control program. But a result of this combination of factors is that more moisture is held in the soil; this somewhat increases crop yields, especially if there are periods of continued dry weather. More importantly, with more soil moisture available, the response of the soil to fertilizers is increased, and it now becomes profitable to use more fertilizer than previously. But the old crop varieties do not respond adequately to the improved soil and fertility conditions; new varieties, that would not have been advantageous under the old conditions, now are much more productive. Perhaps additional new cropping or farming practices are now economical. A whole bundle of new ideas and new farming methods has been adopted, either one at a time or all together. Thus, land management for greater output and greater profitability has replaced soil erosion control as the dominant purpose of the program on this farm. Multiplied by many thousands, this example illustrates what has happened on a national scale.

This shift in the emphasis of programs can be documented in a number of ways. As we have noted, the annual reports of the Soil Conservation Service in 1935, 1936, and 1937, when Bennett was chief, were full of the term "soil erosion," as was Bennett's 1939 book. Although we have not made a detailed count, we hazard the judgment that the word "erosion" appears at least three, and perhaps five, times as often as "conservation"; there are only traces of what we call "management"; and emphasis is upon terraces and other mechanical means of preventing soil loss. In the SCS annual reports for 1938 to 1945, the change in emphasis had begun to appear, although not very explicitly. But in the 1948 annual report, the change was quite explicit, as the following quotations show:

> Modern soil conservation is not directed merely toward maintenance of the status quo. It is dynamic and progressive; it leads to increased and lasting productivity of the land and thereby promotes the common welfare wherever it is practiced. . . . Modern soil conservation does much more than safeguard land. It directly and indirectly results in a wide variety of fundamental benefits. It both increases the yields per acre and lowers the cost of production on most farm land. . . . It results in large savings, such as reduced siltation of streams, ditches, harbors, and costly reservoirs. . . . It helps to alleviate drought damage to crops, pastures, and meadows. It encourages a more flexible and diversified type of agriculture with

a sound physical basis for making annual adjustments in the amounts and kinds of production needed to stabilize the national and world economies. . . .

If there is any activity of mankind that requires the most scrupulous use of all that land science and hydraulics can provide, it is the work of keeping our land permanently productive and making the best use of our water supply. There are some 60 major soil-and-water-conservation measures now being used by the Soil Conservation Service to halt erosion, conserve rainfall, and improve the land. . . . Usually, combinations of several measures are essential, in mutual support of one another, to obtain the most effective conservation. . . . Single-practice soil conservation programs have never been successful on cultivated lands in terms of real and lasting conservation. . . .

Of the approximately 450 million acres now classified as cropland in the United States, about 60 million should be taken out of cultivation altogether, primarily because the land is too steep, too shallow, too poor, too severely eroded, or too susceptible to erosion for successful cultivation. As a replacement for this land, there are something like 100 million acres now in some sort of grass, weed, brush, or tree growth that could be converted to cropland. Much of this would require drainage, as well as clearing or stumping, before it could be used for crops. Other areas are suitable for development by leveling and irrigation. . . . The wisdom of proceeding with the development of these new lands in a sound, orderly way is indicated by the prospects for a considerable increase in the Nation's population during the decades immediately ahead . . . we should finish the development of these lands as rapidly as they are needed, as well as the conservation job on present farm lands by about 1970.[7]

The same point can be made by comparing the *Yearbook* of the Department of Agriculture for 1938 and that for 1957. The latter, entitled simply *Soil*, is described in the editor's preface as an updating of the former, *Soils and Men*.[8] While each is concerned with erosion, conservation, and management of land, and the later one does not fundamentally contradict the 1938 report, the relative emphasis shifts greatly. Whereas erosion, or loss of soil, dominated the first book, the 1957

---

[7] *Report of the Chief of the Soil Conservation Service, 1948*, U.S. Department of Agriculture, pp. 48–51, and 61.

[8] *Soil—The Yearbook of Agriculture, 1957* (Washington: Government Printing Office), and *Soils and Men, op. cit.*

Yearbook is much more concerned with land management for greater output—and, it is hoped, for greater profit. The emphasis in it is upon fertility, soil care, and the like.

The first programs were primarily for the maintenance of the existing basic productive capacity in the land, especially by preventing the loss of soil material through wind or water erosion. While such programs resulted in some increases in productive capacity, this was not their primary emphasis. But the later programs clearly indicate major concern with the building of additional productive capacity and with adding to current inputs as a means of affecting output. This shift in emphasis often made good sense to the farmer. Generally speaking, he was less interested in saving his soil, as such, than in increasing his income. Measures to reduce soil erosion to prevent loss of income at some future date were less appealing than measures to increase his output today or tomorrow. In many cases, small adaptations of erosion control programs led to substantial increases in output.

Similarly, the shift in emphasis made good sense to SCS, primarily because it was a means of interesting farmers in the agency's program and in making them more favorably disposed to the agency. Since SCS was engaged in serious conflict with bureaucratic rivals, especially PMA and the Extension Service throughout this period, it needed to build popular and political support wherever and however it could. Adapting its program to what farmers were interested in was surely one effective device. Presumably, SCS advocated only programs in whose technical soundness it strongly believed; its emphasis upon planning for the whole farm, which often led to controversy with other agencies and farmers, seems proof of this. But, within the range of technically sound programs, a public agency is often wise to push popular programs; in this way, it not only assures its own health and continued existence, but obtains the means for carrying out later programs which currently seem less popular.

But this shift in emphasis of the SCS program is much more dubious from a national or social viewpoint. To the extent that it was effective on the lands to which it was applied—and we must assume that it was effective to a considerable degree—it surely increased total agricultural output of those lands over what it otherwise would have been. Except for the war years, these were years when the national agricultural program was concerned with limiting total agricultural output to meet effective demand at politically acceptable prices. Various expensive programs were being directed to this end. Whatever may have been the public statements of the secretaries of agriculture during this period, a fundamental conflict in purpose and in results of programs existed. One part of the Department of Agriculture was spending large sums of public money to control output; other parts were spending smaller, but still substantial, sums to increase it—and no small part of the rationale for the latter expenditures was the need for public support in the continued struggle of SCS for existence.

Although a conflict has existed between the conservation program and several other programs of the Department of Agriculture during the period since 1933, it is essential that this conflict be kept in reasonable perspective. First of all, a farmer was interested in soil conservation only if his net income increased or remained constant while his soil losses were reduced. If he could expect no gain from a conservation program, he would not enter it. While it is possible that the result of the program might have been to lower his costs, and thus increase his net income, the opposite nearly always occurred. Costs rose as conservation measures were carried out or greater annual inputs were introduced, rather than fell. Thus output and individual farm income had to rise in order to produce an increased net farm income.

If the nation was to have more soil conservation, it probably had to accept more agricultural output. This might have been justified on a regional or national basis had some land been taken out of production, or shifted to a less intensive use, as a direct result of increased output on some other land. In fact, while farmers took some poor land out of production during these years, there is little reason to think that they did so because of soil conservation efforts in other areas. If one assumes that certain land will be farmed no matter what programs are undertaken, it follows that some social gain comes from programs which increase output from the same volume of inputs. This line of argument has been advanced, for example, to justify the inclusion of drainage programs under the name of soil conservation. But these arguments at best tend to excuse, not support, the output-increasing aspects of soil conservation programs.

The activities of the Soil Conservation Service and the Agricultural Conservation Program have been unfairly singled out for criticism because of their output-increasing aspects. In practice, very nearly the whole research program of the Department of Agriculture and of the land-grant colleges increases output. These research programs are supported entirely from public funds; farmers are encouraged to use the results without charge. Similarly, the irrigation programs of the Bureau of Reclamation of the Department of Interior and the flood protection programs of the Army Corps of Engineers and other federal agencies also increase output, and a considerable part of the costs are borne by the general taxpayer. Each of these programs has its supporters, and each tries to rationalize its actions in terms of the national welfare. But the fact remains: some public programs increase agricultural output while other (more expensive) programs reduce or control output.

How might this divergence in purpose be resolved? In the classic free market, the prices of agricultural commodities would be allowed to fall, so as to increase consumption,[9] reduce production, or both. But the reduction of agricultural output resulting from lower prices is slow, costly, painful; the nation, through its political processes, has rejected this approach. This country is not willing to let prices of farm commodities fall low enough to reduce production by significant amounts. Instead, the government has tried to stimulate the removal of land from agricultural production by means of quotas for some crops, payments for voluntary reductions in crop acreage for others, and the like. But inputs of other materials, including fertilizer and machinery, have not been regulated. Theoretically, one could regulate or limit supply to the level where the selling price of commodities would exactly equal parity or some other desired prices. In fact, public programs for the past thirty years have been a curious mixture of output-increasing and output-restricting activities, with no real reconciliation of their purposes and terms. The outlook is for a continuation of the same—at least for some years.

This shift in emphasis from soil erosion to land management in the conservation program had neither a sharply dated beginning or sharply dated conclusion. Changes in the philosophy and policies of the soil conservation program were gradual, but their cumulative effect had produced a marked change by 1950. More importantly, the new program as we have described it, has continued and become intensified to the present. The soil conservation effort of the SCS and the other agencies of the U.S. Department of Agriculture has production-increasing effects to the extent that it is effective at all. In 1950, it was still possible to believe that crop surpluses were temporary and that within a few years much additional output would be needed.[10] By 1964, this clearly was not true; the Department's official position had shifted, as had the consensus of professional opinion.[11] The conflict

[9] With the low price elasticity for most farm products, this would have little effect on consumption.

[10] *The Fifth Plate*, U.S. Department of Agriculture, January 1951.

[11] *Land and Water Resources—A Policy Guide*, U.S. Department of Agriculture, May 1962 (slightly revised September 1962).

between stepping on the gas and stepping on the brake simultaneously had become sharper.

The implications of this situation for the future will be discussed in Chapter 12. At this point, we need only note that the existence of this conflict does not, in itself, demand total abandonment of either soil conservation or production adjustment programs. It does require that each program be critically re-examined for its suitability to the world of the present and of the future. It also suggests that in the budgetary process the difference between practices which increase output and those which maintain the soil be as clearly distinguished as possible so that scarce public funds may be made available only for investment in soil maintenance.

The Small Watershed Program

The evolution of soil conservation work has also been influenced by special legislation which has placed greater stress upon flood control and other watershed management activities. This has been especially true of the work carried on by the Soil Conservation Service.

Under the Flood Control Act of 1936, the Department of Agriculture had been given certain responsibilities, particularly for upstream or small stream problems, to complement the programs of the Corps of Engineers on the larger rivers. Within the Department, responsibility was shared at first by the Forest Service, Soil Conservation Service, and Bureau of Agricultural Economics (BAE); later, BAE was relieved of responsibility for this program. A considerable number of rather intensive surveys were undertaken. However, there was far from complete agreement within and between federal agencies as to the physical effectiveness of various proposed measures—and still less agreement as to the economic feasibility of some proposals. The role in flood retardation of forests and other vegetation, and of measures for upland treatment generally, has a history of controversy and argument. Detailed reports were submitted by SCS to the Congress, which approved work on some eleven watersheds in late 1944.[12] Money was appropriated and programs were begun on these watersheds shortly after World War II.

However, it became apparent to Bennett and others in SCS that this was going to be a very slow and difficult way to get a large watershed

[12] *Report of the Chief of the Soil Conservation Service, 1950,* U.S. Department of Agriculture.

management program under way. Bennett had always stressed the importance of soil conservation as a means of reducing the severity of floods, and especially of lessening the movement of mud and other sediment into reservoirs, channels, harbors, ditches, and the like. From the start, SCS and the Corps of Engineers had disagreed about the efficacy of various flood programs—a controversy to which we shall refer later—but soon after the end of the war, SCS recognized that it could not really touch the problem of the major downstream flood damages. Aside from the technical difficulties of markedly influencing such flood areas by upstream works, the Corps of Engineers was too firmly established in this kind of program. However, flood damages often occurred on small streams, or on smaller tributaries of large streams, where the Corps usually was not interested in trying to plan a project. Although such drainages were smaller than those with which the Corps was primarily concerned, they were typically too large to be handled by single farmers, or even small groups. It was to these areas that SCS looked for its major water management program. Moreover, the SCS personnel gave lip service, at least, to multiple purpose water development, not simply to flood control.

The period from 1944 to 1954 was one of increasingly active rivalry between the SCS and the Corps over the development of a program for these upstream areas. The flood control program as developed through the Corps of Engineers for many years provided substantial windfalls to landowners in areas of flood hazard. Local units of government or individual landowners were required to provide rights of way for dams, levees, and other structures, and sometimes to agree to maintain the works which were built—an obligation that often did not amount to much for many years, when it could be conveniently forgotten. In return, substantial sums of federal money were spent in the area, with economic effect during the construction period, but with more lasting effect in terms of greater land productivity and higher land value. Under the 1936 Flood Control Act, the total benefits, to whomsoever they should accrue, had to exceed total costs; in practice, this requirement was often interpreted generously. This type of generous program had always benefited some rural landowners, especially along the Mississippi and other major streams. Its appeal was obvious, and after World War II, the Corps took steps to take advantage of it. Equally aware of the appeal of such a program, SCS and the new National Association of Soil Conservation Districts (NASCD) also tried to gain some of this rural good will through such programs. Sharp political struggles ensued, which the congressional friends of SCS helped to win.

The result of this conflict was the Watershed Protection and Flood Prevention Act of 1954, which since has been amended several times. In general, the amendments have liberalized the original terms. We shall not try to trace the evolution of this Act, but by 1964 the program was limited to watersheds no larger than 250,000 acres, or slightly more than 10 square miles; and reservoirs not to exceed a capacity of 25,000 acre-feet, with a limit of 5,000 acre-feet of storage which can be allocated for flood control.[13] While these are relatively small areas, the nation is largely made up of many small watersheds, and in fact by far the greater part of the total land area ultimately can be included in spite of the limitations of the 1954 Act. The legislation requires that an application be received from some local organization; in fact, most of them have come from soil conservation districts. Applications must be approved by a state agency, one so designated by the governor of the state concerned—in a great many states, the agency is the state's soil and water conservation committee. Projects approved by the state agency must also be approved by the SCS in Washington, if they exceed a cost of $250,000 or have a single structure with a capacity of over 2,500 acre-feet; and the larger projects also must be approved by the congressional committees most directly concerned.

The really significant feature of this program for small watersheds is the provision for cost-sharing. Some local group with adequate legal powers to raise the necessary funds, or the state, must assume the non-federal share of the costs. These costs include land treatments on non-federal land in the watershed; acquisition of necessary rights of way (except that the local group pays only half the right-of-way costs for public recreation or fish and wildlife development); at least 50 per cent of the construction costs allocated to agricultural water management and public recreation; all of the costs for municipal and industrial water supplies and some other miscellaneous purposes; and some other incidental costs, including some maintenance costs. The federal government bears all costs allocated to flood prevention; half of the costs allocated to recreation, wildlife, and agricultural water management; and some other miscellaneous charges. Moreover, the federal government makes loans available to help local groups raise their share of construction costs.

---

[13] Legislation is pending before Congress (1965) which would permit the allocation of up to 12,500 acre-feet of storage for flood control.

With such generous bearing of costs for flood control by the federal government, it is not surprising that up until 1962, when recreation was added as a proper purpose, the overwhelming emphasis of the program was upon flood control or prevention. In fact, many students have felt that significant opportunities for multiple purpose water management were missed because local groups were so intent on getting free, or nearly free, federal money for flood prevention. With aid available on such generous terms, there was naturally a rush of projects, faster than SCS could deal with at first. By April 1, 1964, the SCS program for small watersheds had 100 completed projects; 447 authorized for operations, but not completed; 416 authorized for planning, but not for operations; and it had received, but not processed, 1,125 applications for projects—or a total of 2,088 projects at some stage of consideration. The projects completed and those authorized for operations but not completed affected over 31 million acres; those authorized for planning, almost 33.5 million additional acres; and those for which applications were still pending, over 84.5 million acres—making a total for all categories of about 150 million acres.

Although the Small Watershed Program is national, its operations have tended to be highly regionalized. More than half of the total number of projects, total area, and total funds expended have been in Texas and Oklahoma. Since this is a region of serious flood hazard and a multitude of relatively small river basins, it is an area in which the Small Watershed Program works well. More than a third of all the land in Oklahoma is included in watersheds which have applied for assistance under this program.

In 1962, during one of the times when the 1954 Act was being liberalized, recreation was added as one major objective. This arose partly from the Department of Agriculture's growing interest in outdoor recreation as a desirable use of land and water in the modern period. But it also reflects the growing popularity of the idea of recreation in Congress and among the general public. The relatively small reservoirs built under this program often have significant recreational potential. By law, developments eligible for cost-sharing must be open to the public for recreational use; but it does not preclude the charging of a reasonable fee for use of such areas and facilities.

In terms of our definition and analysis of soil conservation, the Small Watershed Program partly maintains basic soil productivity against further deterioration caused by repeated flooding. But, in larger part, it is a program for adding new basic productive capacity to the land. Land subject to frequent overflow is severely limited in economic uses;

the same land, if the flood hazard is removed or greatly reduced, can be used much more intensively. Its basic productivity has been increased; so, usually, has its value per acre. It is precisely for reasons of higher values that local interest in the Small Watershed Program is so great. To the landowner, who can get the value of his land increased by public expenditure, the program obviously makes sense, and he naturally seeks to get a project for his area. More intensive use of the land follows naturally, and its total agricultural output rises.

The program has also made sense for SCS, given the political milieu in which it has had to operate. This agency could hardly expect the support of farmers in its struggles against Corps encroachment into rural areas, when the Corps could offer so much more favorable terms. In self-defense, therefore, SCS embraced the small watersheds idea. But it was perhaps more than a defensive weapon; it also enabled SCS to build support among farmers for the whole range of conservation programs. One can believe that SCS was sincerely convinced of the wisdom of flood prevention programs from the point of view of land management, and, at the same time, it realized how well this program fitted into its political program.

From the national viewpoint, however, the same questions must be raised here that were raised about the whole change in emphasis toward land management and away from erosion control. The nation did not need the added productive capacity developed through small watershed projects, any more than it needed additional capacity developed in other ways. There was the same conflict within the Department of Agriculture between programs which increased basic productive capacity and those which limited current output. Though one may criticize the Small Watershed Program in general, one must also admit that it was generally no worse, and in some respects was better, than the flood prevention programs of the Corps of Engineers. As long as the latter exist, SCS and its supporters must seek to gain something equally good for their program; no political realist could expect anything else.

The Great Plains Conservation Program

Another soil conservation program, also with specific legislative approval, is the Great Plains Conservation Program established in 1956. Although limited by law to one major agricultural region of the United States, this program has several features which could be enormously significant if adopted nationally. It provides for a longer-range program for conservation and land-use adjustment than had previously been possible, and it is additional to other federal conservation programs.

The Great Plains includes parts of ten states, but none lies wholly in the Great Plains. Federal programs of all kinds have been particularly pervasive in the Plains. As the nation developed, Indians from areas east of the Mississippi were pushed westward into the Plains, and eventually confined to special limited reservations—opening the land to white settlement. Individuals acquired the land, all originally owned by the federal government, in a variety of ways—by direct purchase; by homesteading; by purchase from railroads which had received major aid from the federal government in the form of land grants; and in other ways. The pattern of settlement was dictated largely by the method of land survey, subdivision, and sale. Railroads were built across the Great Plains primarily as a means of reaching the Pacific Coast, but they opened the region to economic development.

Several basic difficulties have plagued the Great Plains. The climate is variable; more significantly, the variation takes place around the point which is critical to crop production. In a good rainfall year, bumper crops can be produced; in a severe drought year, no crops, or at least unprofitable ones, are produced. Other regions of the United States have climatic variations, possibly as severe as those of the Plains, but none has its variability so directly around a critical point.

The federal government has been responsive to the plight of the farmers in the area. In the first World War and afterward, "feed and seed loans" were extended widely to settlers in the Plains. In fact, while called loans, many were not repaid, nor was it intended that they should be. The conditions of the loans were far different from those a commercial lender would have imposed. In the depression of the thirties, the Great Plains received massive federal aid in the form of direct relief, public works of many kinds, purchase of cattle during severe drought periods at prices far above the market, loans to operators of small farms, and in other ways. In more recent times, large-scale federal expenditures for water development on the Missouri River and its tributaries by the Bureau of Reclamation of the Interior Department and the Corps of Engineers have flowed into the Plains.

The form of settlement, and more particularly the size of farms, established during the development period in the Great Plains was not suitable for continued use of the available resources. It is fashionable in some quarters to blame the inadequate size of the original farms upon the Homestead Act, which established 160 acres as the farm unit. While this was highly influential, parcels of land sold by the railroads were subject to no acreage limit, but were in sizes sought by farmers, and these also were predominantly in 160-acre tracts. Moreover, in the systems of local government, roads, schools, and other public services, the pattern of more humid areas to the east was transplanted to the Plains. Possibly more than any other agricultural region of the United States, the Great Plains area has been the recipient—one might almost say, the victim—of technological and social change; adjustment has been continuous but has never caught up with the underlying causes of change. Finally, wheat—the traditional Plains crop—is subject to a severely inelastic demand, but a demand which has moved up and down over wide ranges during war and depression.

The result of these and other factors has been a very unstable agriculture in the Plains. At one time or another, much land has been plowed which cannot be kept permanently in crops because of erosion, primarily from wind, or which would not be profitable in crops over a long period. Plowing has occurred during the settlement waves, and during each world war; left alone, this land has reverted to weeds, and ultimately to grass. The Soil Conservation Service has estimated in recent years that 14 million acres of land in the Plains were used for crops that could not be kept there permanently because of the erosion hazard; another 10 million acres, according to an estimate of Mont H. Saunderson, are unprofitable for production of wheat on a long-term basis.[14] One major problem in recent years has been caused by the federal price supports for wheat, and the value arising out of acreage allotments for wheat (which every farmer was loath to risk losing). Land for wheat production has acquired a value which it is impossible to sustain if the land is used for grazing. In addition, the costs of reseeding grass for quicker recovery are very high relative to the value of the improved grazing land. The public programs, by creating impossible hurdles to economic adjustment, become the rationale for their own continuance.

The Great Plains in the past, also, has had an unstable farm population. While every locality has some farmers who have remained on the same land over long periods, an unusually high turnover of farms has occurred—not only in times of adversity, such as drought, but also in

[14] Soil Conservation Service, *Facts about Wind Erosion and Dust Storms on the Great Plains*, Leaflet 394, U.S. Department of Agriculture, June 1955; Saunderson, "Range Problems of Marginal Farm Lands," *Journal of Range Management*, Vol. 5, No. 1 (January 1952).

times of prosperity and bumper crops. A high proportion of the farmers have had only relatively short experience in the Great Plains—experience which may have included only years which are not typical in the long run. There is also reason to believe that a high proportion of the farm operators have not looked upon the Plains as their long-time home, but rather primarily as a place to make some quick money, before moving to a more attractive place in which to live. The natural conditions of the area have been difficult enough. But the human factors have made it even harder to work out long-term resource use programs adapted to the area.

Some part of the Plains has a drought nearly every year; droughts were especially bad in the northern Plains in the thirties and in the southern Plains in the fifties. The acreage damaged by wind erosion in the Great Plains has varied greatly from year to year (Figure 7), depending in part upon the severity of windstorms and in part upon the area of sandy and other easily erodible soils plowed annually. Wind damage has been most severe in the central and southern parts of the Great Plains, although it has also occurred in the northern part. Two extensive areas—one in southwestern Kansas and southeastern Colorado, and another in east central New Mexico and west Texas—have had more than 50 per cent of the cropland damaged by wind erosion. (See Figure 8.)

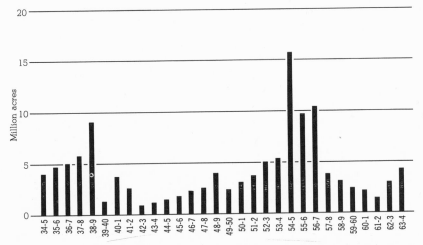

Figure 7. Acres of land damaged annually in Great Plains, seasons 1934–35 to 1963–64 inclusive. (Chart from Soil Conservation Service, U.S. Department of Agriculture.)

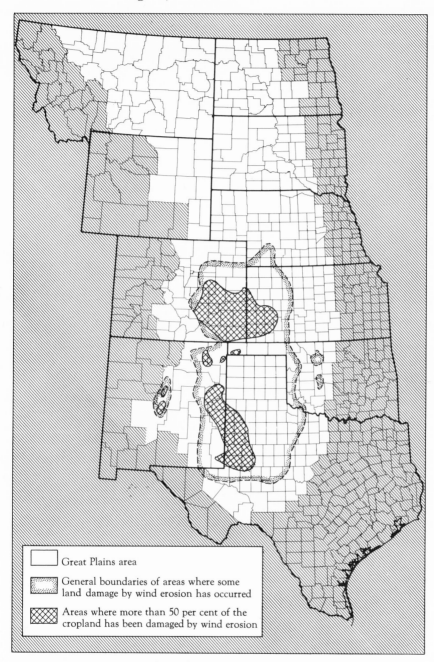

Figure 8. Great Plains area where wind erosion has been severe. (April 1, 1954 map from Soil Conservation Service, U.S. Department of Agriculture.)

The Great Plains Conservation Program provides for contracts between the federal government and the farmers over periods of three to ten years. Farmers are free to refuse to participate of course, but when they do enter into a contract it must be based upon a conservation plan for the whole farm or ranch. The contract covers the full range of conservation practices included in the plan, and must be for the full period of the plan. This differs from the regular SCS and PMA programs under which the farmer may choose some practices only, or may include only part of his farm. The objective of the Plains program is evident: to work out and put into effect a long-range conservation program for whole farms, which it is hoped the farmer will continue after the initial contract expires. By attacking the conservation and the production adjustment problems jointly and over a longer period, it is hoped that more basic and continued adjustments will be secured than have resulted from earlier programs.

Up to mid-1964, over 16,000 contracts had been signed between the federal government and farmers under the Great Plains Conservation Program (Table 3). In addition, some 4,700 applications were in process. The number of signed contracts is only about 5 per cent of the total number of farmers who were in this region at the time of the 1959 Census of Agriculture. Thus the program has had a limited effectiveness because of low participation.

TABLE 3. Status of Great Plains Conservation Program, June 30, 1964[a]

State	Great Plains contracts signed		Cost-share obligations of federal government ($1,000)	Planned cropland conversion (1,000 acres)	Unserviced applications (number)
	(number)	(1,000 acres)			
Colorado	937	3,345	6,069	85	516
Kansas	1,665	1,402	4,319	58	529
Montana	774	3,423	4,102	116	313
Nebraska	2,202	2,826	6,710	140	802
New Mexico	736	5,861	4,812	46	236
North Dakota	1,842	2,579	4,379	229	545
Oklahoma	1,246	1,836	4,102	62	333
South Dakota	688	2,443	3,724	93	208
Texas	5,813	8,937	18,261	272	1,179
Wyoming	218	1,231	1,876	13	71
Total	16,121	33,883	58,353[b]	1,112	4,732

[a] Tentative totals.
[b] Cumulative expenditures about $36 million, leaving an unspent balance of $22 million.
SOURCE: Based on processed tabulation of the Soil Conservation Service, U.S. Department of Agriculture, September 23, 1964.

The signed contracts include about 34 million acres of land, which is less than 10 per cent of the total area of the Plains, a somewhat larger proportion of land in farms, and somewhat more than one-third of the area of cropland harvested. However, one may assume that the contract acreage includes far more than cropland. By the end of fiscal 1964, the federal government had obligated about $58 million under the Great Plains Conservation Program; however, actual expenditures as of that date were about $36 million. These sums in the context of the total cost of the various programs are relatively small.

The program as it had evolved to mid-1964 included the planned (but presumably yet not fully attained) conversion to other uses of around 1 million acres of cropland, probably most of them for grazing. This is clearly a step in the right direction, but only a small step compared to the SCS estimate of 14 million acres unsuited for continued cropping, or to 24 million acres if one includes Saunderson's estimate of the additional area of economically submarginal land in the Great Plains.[15]

The various accomplishments of the program up to mid-1963 are shown in Table 4. All of them have some "conservation" content, but some are more valuable for increasing basic productive capacity than for maintaining such existing capacity. Establishing permanent vegetative cover, if on land now or recently growing a crop, is a means of preventing wind erosion; so is the establishment of windbreaks. But the control of shrubs, the development of livestock water by various means, the leveling of irrigated land, and some other practices are more for the creation of additional productive capacity than for maintenance of that which now exists. A rough classification of the various practices suggests that one-third to one-half of total expenditures is to maintain basic productive capacity and that one-half to two-thirds is to increase capacity. Obviously, there will be borderline cases.

The Great Plains Conservation Program, to summarize, represents a definite forward step in both concept and results toward more permanent conservation adjustment; but its rate of progress in proportion to the problem has been very slow. No doubt this has been hampered to some extent by the fact that price supports and other programs for the area's crops have tended to make conservation farming less profitable than a more exploitative type.

[15] *Op. cit.*

TABLE 4. Conservation practices of the Great Plains Conservation Program, 1956 to June 30, 1963

Num-ber	Conservation practice	Unit	Extent to which applied	Cost-share paid by federal government		Aver-age unit cost paid
				Amount ($1,000)	Per cent of total	
GP– 1	Permanent vegetative cover	1,000 acres	486	$3,691	13.4	$ 7.59
GP– 2	Field strip cropping	1,000 acres	153	211	0.8	1.38
GP– 3	Contour strip cropping	1,000 acres	58	247	0.9	4.24
GP– 4	Contour farming	1,000 acres	1	2	a	1.56
GP– 5	Reseeding range land	1,000 acres	443	2,614	9.5	5.90
GP– 6	Windbreaks	1,000 acres	10	414	1.5	40.38
GP– 7	Permanent sod waterways	1,000 acres	7	613	2.2	81.82
GP– 8	Terrace construction	1,000 miles	15	2,419	8.8	156.98
GP– 9	Diversion terraces, etc.	1,000 miles	2	516	1.9	292.62
GP–10	Furrowing, etc.	1,000 acres	91	110	0.4	1.21
GP–11	Dams for erosion control	1,000	8	1,262	4.6	155.27
GP–12	Channel lining, etc.	Number	482	152	0.6	316.23
GP–13	Streambank protection	1,000 feet	54	33	0.1	.62
GP–14	Spreader ditches, etc.	1,000 acres	40	775	2.8	19.50
GP–15	Reorganize irrigation systems	Number	1,004	1,895	6.9	1,887.37
GP–16	Leveling land	1,000 acres	43	1,494	5.4	35.04
GP–17	Dams, pits, ponds, etc.	Number	135	165	0.6	1,222.00
GP–18	Lining irrigation ditches	1,000 feet	491	439	1.6	.89
GP–19	Wells for livestock water	Number	3,472	1,908	6.9	549.57
GP–20	Developing springs	Number	296	64	0.2	214.79
GP–21	Dams, etc. (vegetative cover)	Number	6,367	3,286	11.9	516.12
GP–22	Pipelines for livestock	Miles	602	602	2.2	999.84
GP–23	Controlling shrubs	1,000 acres	1,450	3,756	13.7	2.59
GP–24	Constructing permanent fences	Miles	2,558	854	3.1	333.80
	Total			$27,522	100.0	

a Less than 0.5 per cent.
SOURCE: Based on mimeographed tabulation provided by the Soil Conservation Service, U.S. Department of Agriculture, May 1, 1964.

Approximate Expenditures for Soil Conservation

The several public and private programs for soil conservation have not been costless. On the contrary, substantial sums have been spent for this purpose over the past thirty years or more. Because the term soil conservation is neither easily and directly defined and there is no complete agreement as to the items to be included, it is difficult to state exactly how much has been spent for this purpose. Moreover, there is not complete information about expenditures other than those of the federal government.

Omitting the much more expensive programs for crop storage and price supports which conceivably could be called conservation, those federal programs more or less directly concerned with soil conservation have cost more than $9 billion over a thirty-year period, as Table 5 shows. More than half of this total ($5.8 billion) was for the Agricultural Conservation Program of the ASCS, which we have described only briefly in this chapter pending a review of its accomplishments in Chapter 8. This could with equal accuracy be called a farm income, or even farm relief, measure, since recipients of federal payments do "earn" them by various "conservation" practices; but throughout there has been emphasis upon sharing these payments rather widely. Another sizable chunk ($1.6 billion) is for the Conservation Reserve of the Soil Bank; this is in large measure a production adjustment program, but it does have important consequences for conservation. About $1.4 billion has been spent on the program of the Soil Conservation Service, which is almost wholly a soil conservation expenditure under almost any definition of the term.

TABLE 5. Approximate expenditures for soil conservation, 1934–1964[a]

Federal program	Total expenditures, fiscal years 1935–1963 ($ billion)	Average annual expenditure, fiscal years 1959–1963 ($ million)
By ASCS:		
Agricultural Conservation Program	5.8	239
Soil Bank	1.6	295
By SCS:		
Soil Conservation Service[b]	1.4	95
Watershed protection and flood prevention	.4	54
Total	9.2	83

[a] Includes only programs officially designated as soil conservation. Excluded are: federal programs on federally owned land and research programs on soil conservation; flood control and other programs of the Army Corps of Engineers, Bureau of Reclamation of the U.S. Department of Interior, and Tennessee Valley Authority; and agricultural programs designed to raise and stabilize farm income and to adjust agricultural production.

[b] State contributions to the soil conservation district programs for one year only (as reported by the Soil Conservation Service to the House Committee on Appropriations, in *Hearings on Fiscal 1964 Appropriations*, p. 1027) by use of funds in 1963 were:

State committees	$ 1,503,057
Assistance to districts	2,806,563
Watershed:	
Investigations and planning	1,806,268
Works of improvement	5,253,493
Other	516,590
Total	$11,885,971

SOURCES: Office of Budget and Finance of the U.S. Department of Agriculture and U.S. Treasury Department, "Combined Statement of Receipts, Expenditures and Balances of the United States Government," fiscal years 1959–1963.

Moreover, the rate of expenditures on these programs has been rising. Had expenditures for the whole period been at the average rate of the last five years, the total would have been almost $20 billion, instead of $9 billion. Some of the apparent rise over the long period has been caused by rising prices. But the rate of expenditure has also risen within the five years averaged in Table 5, from somewhat more than $500 million at the beginning to over $700 million annually at the end, during a period when the general price level was relatively steady.

We do not intend at this point to comment on whether or not these expenditures were worth their cost in terms of results obtained, or whether the money spent was used in the most efficient way—that is, to buy the most soil conservation. The sums of money are large, even by contemporary U.S. standards, and yet there can be no serious question that the nation can afford expenditures of this magnitude if they are necessary.

One can raise the question, however, whether the use of funds of this amount was for the best national program. Without considering other demands for federal funds outside of agriculture, would these same funds have produced more benefits to farm people and for the nation if invested in human, as contrasted with soil, resources? Rural areas have typically had inferior schools, libraries, hospitals, and social services of all kinds. Federal aid amounting to many millions of dollars would surely have had a major impact if used to improve these services. The results of investment in people show up slowly, it is true; but the soil conservation program has now extended over a full generation. For the future, will the national interest be served best by a continuation of past programs or by some shift in emphasis?

Summary

From a review of the evolving soil conservation program in general, that of the federal government more directly, and that of the Soil Conservation Service in particular, a few general conclusions emerge:

1. The program of the Soil Conservation Service has been pragmatic; the agency has sought to solve the soil conservation problems as it found them. It has sought to help farmers. Although it did not always agree with farmers, it has used its own judgment, and it has tried to win farmers over to SCS viewpoints; help to farmers has always been high in its list of objectives. In part, by thus serving

farmers, SCS has been motivated by objectives of public interest; but this service has also been a most effective way of winning farmers' support—and, thus, the political support which was necessary for the survival and health of the agency. Without widespread support, SCS would have had severe difficulties in getting appropriations and perhaps even in continuing its existence.

2. As part of this effort to serve farmers, the SCS program, by meeting local needs, has varied by regions and states. If the problem was erosion of rolling cropland, then terraces and other devices were employed; if the problem was drainage of wetlands, then drainage programs were carried out; and so on. The agency never seems to have felt free to attack severe erosion problems only in those regions or localities where they existed; a program with something for every area, and if possible for every farmer, was politically if not technically necessary.

3. The objectives and nature of the federal soil conservation program have been confused by the establishment of an essentially income-raising program with incidental conservation benefits under the name of soil conservation. Two parallel programs in the Department of Agriculture, with similar names but significantly different specific practices and methods of operation, have tended to blur public understanding of the basic conservation problem. Rivalry between these programs has surely affected the content and operations of each.

4. The SCS program has changed materially since it was begun in the early thirties. New programs, such as those dealing with small watersheds and the Great Plains, were added with specific congressional authorization. More important, but less obvious, have been the great changes in emphasis from that of the original soil conservation program. Begun originally to control erosion, or to maintain the existing basic productive capacity of the land, it has gradually shifted to what is called land management. This includes not only the maintenance of existing basic productive capacity but also the building of new capacity and the encouragement of increased annual inputs into existing production functions. This major shift in emphasis did not have direct congressional authorization in the form of new legislation; it had, however, indirect congressional approval in the form of continued and increased annual appropriations for programs.

Change in public programs is not necessarily bad—in fact, may be highly desirable. Programs which remain the same in spite of major changes in social, economic, technological, and other conditions rightly are often roundly condemned. The shift in emphasis of the various soil conservation programs does not warrant criticism merely because it represents a change in the original objective. Rather, it can be criticized because the fundamental changes in program that have occurred have not been adequately debated and considered by the nation at large. However much the present soil conservation programs may be defended by those who carry them out, or by top officials of the Department of Agriculture, a fundamental conflict exists among the Department's programs aimed at increasing agricultural productive capacity and output, and its programs aimed at controlling output to meet market demands. This conflict among programs is not limited to soil conservation. One need not argue for termination of soil conservation or of output-controlling programs, or both, in order to make the point that each needs fundamental re-examination.

Chapter 5

SOIL CONSERVATION AND RAPID AGRICULTURAL CHANGE

A MAJOR part of the setting for soil conservation over the past generation has been the concurrent change occurring throughout agriculture. The U.S. economy and society are dynamic and changing; such change is expected and desired in the United States to a degree few societies of the world have ever experienced in the past. This broad generalization applies to agriculture also. But the past thirty years or so have been marked by such extremely rapid agricultural changes in the United States that they can fairly be called revolutionary.

The broad changes in farm population, agricultural output, farm income, and other variables have been largely, but not wholly, independent of the soil conservation program of the same three decades. That is, had there been no organized public soil conservation programs in agriculture most of the other changes would have occurred, and to more or less the same degree. These other changes have had, by and large, more effect upon soil conservation efforts than soil conservation alone would have had on other agricultural developments. Although these broad changes in agriculture have had some impact upon soil conservation during the past thirty years, their full effect will be felt only in the next decade or more. And during those coming years, there will almost certainly be other major changes in agriculture generally—changes whose full impact will be felt upon soil conservation only in the still more distant future. In such periods various forces may tend toward equilibrium, but rarely reach it because change occurs too fast. We shall return, in Chapter 11, to the probable nature of future changes, both in agriculture and in the economy and society as a whole; and in Chapter 12, to their probable effect upon soil conservation programs. But here we review only past developments.

This chapter is concerned primarily with the economics and the goals of natural resource use. Changes in technology also enter indirectly, in terms of their effects upon inputs and outputs. With a drastically changing total economy, the economics of both agriculture as a whole and the individual farm have changed, as have the goals of agriculture. In one way, farmers have become more economics-minded or profit-minded—agricultural production now is more for the market, less for family consumption; more inputs are purchased; and net farm income has become more sensitive to changes in market prices. In another way, farmers have become less profit-minded; they are now more aware of the long-term effects on their land of soil erosion, and more concerned to preserve their land. In this latter sense, the role of the Soil Conservation Service (SCS) and its leaders has surely been a major factor.

Many agricultural changes we describe have their counterpart in shifts in the role of natural resources generally. As industry and the general economy have developed, more resources have been required, but the relative dependence upon raw natural resources is less. That is, given amounts of natural resources are processed more, and the service sector of the economy is multiplied, so that a given quantity of natural resources today supports a larger total economic output than once was the case. This does not mean that natural resources are any less essential than they ever were, but only that more sophisticated and complex uses are made of the resources.

Changing Farm Population and Number of Farms

From earliest colonial times down to about 1920, the farm population grew steadily as the course of settlement pushed westward and farmers settled on suitable lands in all regions. As a part of this settlement process, new land was cleared of forest or otherwise drawn into production, and organized into additional farms. During this same long expansion period, the numbers of work horses and mules on farms also increased, for this was the era of animal power on farms. The number of people on farms, area of farm land, and number of work horses and mules expanded at more or less the same rate. (See Figure 9.) In other words, the inputs of labor, land, and power increased more or less proportionately.

The farm population reached its peak in 1916 with 32.5 million people, according to the series on farm population which was used for

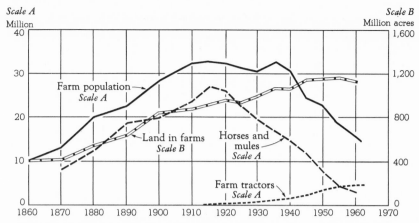

Figure 9. Land in farms, farm population, horses and mules on farms, and farm tractors, from 1860.

several decades. From 1910, when the first data on an annual basis were available, until 1941 farm population was about 30 million. During these three decades, total farm population varied only between 30 million and 32.5 million, or was relatively constant. This relative stability maintained, in spite of high birth rates in rural areas, only because many farm people migrated to cities each year.

During World War II, the rate of migration from farms to cities increased sharply, and has remained high ever since. For these past twenty years or more, the rate has been high enough to cause actual declines in farm population. Many persons who continued to live on their old farms no longer continued to operate them as real farms, and the definition of farm population has been adjusted to reflect this change. By the old definition of farm population, the total number was down to 20.5 million by 1960; by the new definition of farm population, it was down to 15.6 million in the same year—a cut of almost exactly half in twenty years. Whatever may be the best definition and the exact relationship, truly dramatic changes clearly have occurred. Moreover, they have been widespread. During the fifties, more than half of all the counties in the United States lost population at a time when the nation was gaining by 19 per cent, and the rural areas of many other counties also lost. In a good many cases, the loss in the fifties, while larger than previously, was only a continuation of earlier trends toward fewer people in rural areas.

These changes in total farm population arise out of, or cause, or occur simultaneously with, proportionate changes in the number of

operating farms. The number of farms increased steadily throughout
the long period of territorial expansion for agriculture. There have
been some variations in definitions of farms used by the U.S. Bureau of
the Census, and in their application in the field, but the broad trends
are evident from the census data. According to the censuses taken at
10-year intervals, the peak number of farms was in 1910, with 6.36
million; the agricultural censuses taken at 5-year intervals place the
peak in 1935, with 6.81 million farms. Some movement back to the land
took place during the severe depression years of the early thirties, but
there is also reason to believe that the enumeration of very small farms
was more complete during the 1935 census than in others. The number
of persons per farm has been relatively steady: 5.07 in 1910, 5.03 in
1920, 5.03 in 1930, and 5.23 in 1940. Using the old definition of farm
population, the figure had risen to 5.58 by 1960; using the new definition, it had fallen to 4.25.

By any reasonable definition and series of data, the total number of
farms remained quite steady from 1910 to 1940, during the same period
when the total number of farm people was also steady. Major changes
have taken place in the number of farms since 1940, however. In 1960,
3.67 million farms were reported by the Census of Agriculture, or 59
per cent of the peak number shown by the 1910 Census, and 54 per
cent of the peak number shown by the 1935 Census. Although our
comments are limited to the changes taking place up to 1960, there is
every reason to believe that the same trends have continued, and that
the total number of farms has declined further. Changes in the number
of farms, like changes in total farm population, have taken place
throughout most of the country. Only a relatively few areas show increases in the number of farms in recent years, although the rate of
decrease varies somewhat from locality to locality.

The changes in the total number of farms are clearly related to
the age of farmer, and somewhat to the size and type of farm.[1] The
prime cause of the decline in the total number of farms has been the
refusal of young men to enter farming, primarily because of the poor
income prospects. Before actual entrance into farming, a farm youth

[1] See the following articles in the February 1963 issue (Vol. 45, No. 1) of the
Journal of Farm Economics: Jackson V. McElveen, "Farm Numbers, Farm Size and
Farm Income"; Marion Clawson, "Aging Farmers and Agricultural Policy"; G. S.
Tolley and H. W. Hjort, "Age-Mobility and Southern Farmer Skill—Looking Ahead
for Area Development"; and Don Kanel, "Farm Adjustments by Age Groups, North
Central States, 1950–1959."

is relatively footloose; he can and does shift his employment to non-farm occupations and his place of living from farm to nonfarm location. However, once a man has begun farming and has reached about 35 years of age, he is most reluctant to leave. His skills are not transferable, and he fears—often rightly—difficulty in getting an urban job. Negroes from rural areas face job discrimination also. As a result, relatively few men who once begin farming withdraw from it earlier than they would normally retire. The extent of accelerated or above-normal withdrawal of older men from agriculture, in the face of low farm incomes, depends somewhat on the assumptions made. Kanel estimates that as much as 25 per cent of the total decline in farm numbers may result from this factor, and Clawson estimates it at 8 per cent.[2] In any event, a disproportionately large number of farmers today are relatively old, and are on farms which no young man will be willing to take over when the present operator is unable to continue. Further drastic reductions in the number of farms are almost certain. The Clawson estimate shows a range of between 600,000 and a million farms by the year 2000; whatever the exact figure may turn out to be, it is likely to be far below the present number.

The decline in the number of farms and in farm population has meant a reduction by half in the total labor input into agriculture. At the same time, land input changed relatively little; acreage of cropland declined slightly; while acreage of all land in farms rose somewhat (but this increase in acreage resulted largely from counting as farm land acres previously used, but not controlled, by the farmer, especially in the Great Plains). The sources of power on the farm changed completely, from horses and mules to tractors and trucks. The similarity of the change in the number of farmers and of work horses and mules during the expansion period of American agriculture may extend also to the contraction period. The number of horses and mules began to trend downward roughly twenty years before the number of farms began to decline, and has been swifter—the horse has a life span of twenty years, more or less, compared to man's seventy. Both horse and farmer have been the victims of changing farm technology, which no longer required such large numbers of either. When the number of horses and mules began to decline, one of the first signs was an aging of those still in use and a reduction in the number of young ones.

[2] The Kanel and Clawson estimates are in the *Journal of Farm Economics* cited in the preceding note.

The Effect of Settlement Patterns and
Land Layout on Soil Conservation

Ownership of land in the United States has always been in relatively small units in spite of some conspicuous exceptions. If the "family farm" is defined as one including no more land than one or two men (farmer, or farmer and son) can operate with no hired labor, or only limited amounts of seasonal hired labor, then the family farm has not only been a strongly held ideal but also a dominant fact in agriculture throughout U.S. history. The precise acreage depends upon the type of farm and upon agricultural technology. Several thousand acres of grazing land or a few thousand acres of wheat land may conform to the definition as well as a very few acres of tobacco.

 Land titles in the United States have been predominantly fee simple—again, in spite of some exceptions. If ownership of land includes a "bundle" of many different rights, then this nation has tended to give all, or nearly all, such rights to the owner; in some cases, some rights have been split off but much less so than in most other nations of the world. Some proportion of all farms have been operated by tenants throughout the history of the United States, but the relative importance of tenancy has shrunk considerably in the past thirty years or so.

The form of the cadastral survey and of land settlement has varied in different parts of the country, as Figure 10 shows. For the greater part of the nation, however, a rectangular cadastral survey has been employed; property lines run in cardinal directions. The result has been described as "square farming in a round country." Not only do farm property lines run north-south and east-west, but so do most roads, and all the many social services tied to, or arising out of, the roads. Moreover, fields within farms also tend to run parallel to farm boundaries, and rows of crops within fields—even when this means going up and down the hill, rather than around it. In those parts of the country where the rectangular cadaster was not used, its very great advantages in the description of land titles and land records were lost; and it is doubtful that much was gained in terms of better conformance of farm property lines and field layout to the natural topographic features of the land. Those responsible for land surveys and the early settlers generally were not much concerned about the soil conservation problem, but they were much concerned to have their property lines known and identified.

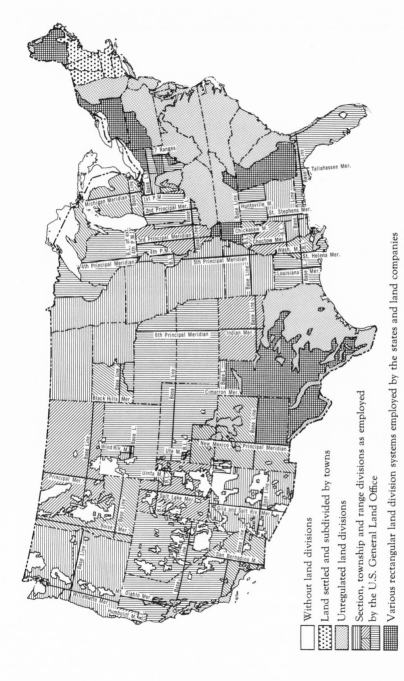

Figure 10. Land division types. (Map from Agricultural Research Service, U.S. Department of Agriculture.)

Without land divisions

Land settled and subdivided by towns

Unregulated land divisions

Section, township and range divisions as employed by the U.S. General Land Office

Various rectangular land division systems employed by the states and land companies

Roads are necessary in rural areas if farmers are to buy needed supplies and to market their output. Rural roads typically have followed farm property lines—not only along section lines in the regions where the rectangular cadastral survey was used, but along other farm property lines elsewhere. Typically, each farmer donated a strip of land half the width of the road right-of-way, as an inducement to public construction of the needed road. If every square-mile tract—"section," in cadastral survey terminology—had a road on each side, there would average two miles of road per square mile of rural territory. As a matter of fact, there are almost exactly 3 million square miles in the forty-eight contiguous states, and almost exactly 3 million miles of roads of all kinds, or an average of a mile of road per section. Extensive mountainous and grazing areas have far fewer roads, of course, while rural areas containing small farms have more roads.

Although acreage of land per farm varies a great deal, both among different regions and among farms within the same region or locality, for the nation as a whole, the average acreage per farm was remarkably stable for a long time. It increased very gradually from 134 acres in 1880 to 151 acres by 1930; since then, it has risen sharply, to over 300 acres. A small part of this rise has resulted from bringing more land into farms, particularly in the Great Plains. There, much land which formerly was not associated with any single farm operation but often grazed indiscriminately, has been brought definitely under control by the farm operator. Most of the increase in average acreage per farm, however, has been caused by the major decline in the number of farms.

The rural settlement pattern in the United States and rural services generally were geared to animal power on the farm and to animal transport on the roads. By and large, this implied farms averaging 160 acres and roads at intervals of one mile in each direction. In the past thirty years, while the sources of farm power and of road transport have been so completely changed that each is now completely mechanized, with tractors and trucks, the average size of farms has been changing. (We shall consider future changes in Chapter 11, but we may note here that farms averaging 800 to 1,000 acres for the whole country are implicit in the general projections of increases in the number of farms.) The nature of rural services also has changed enormously in the past generation, but the pattern of rural roads has changed but little.

The SCS and other public agencies and private organizations interested in promoting soil conservation have had to deal with farmers and farms as they found them. A major part of what the SCS found was square farming, and a major part of its effort has been to introduce round farming—contour terraces, farming on the contour, and other specific practices. Conservation plans had to be developed for each farm as it existed. This often meant that land had to be used more intensively than the SCS technician would have preferred, because the farmer had too little land of good quality to produce a reasonable income. But changes in the number and average size of farms can change this situation drastically. A conservation plan for an 800-acre farm may be very different from the sum of five conservation plans for five 160-acre farms in the same area. Great changes in land use and in conservation within farms will be possible as the size of farms increases.

Accompanying, or as part of, these changes in the number and average size of farms has been a major sorting-out process for agricultural land. By trial and error, farmers—and technicians—have learned which lands were best adapted to agriculture. There has been a great deal of land abandonment in American agricultural history. Much land settlement was ill-advised, from the point of view of the settler's own welfare as well as from its effect upon soil conservation. Farming has gradually been concentrating upon the "better" land; but the definition of "better" and "best" has also changed as technology has changed. The intensive margin of farm operation has been pushed forward considerably on the better lands, even while some of the poorer lands were being abandoned. There is no reason to expect that this process will be reversed.

New Relationships in Agricultural Inputs and Outputs

Change in the combination of productive inputs into the agricultural process has taken place over the past several decades, but it has been most drastic since 1940.[3] (See Figure 11.) Labor and land inputs have shown less change than some others. As we have noted, the input of labor rose to a plateau and then declined greatly after 1940, to about half the height of the plateau. The input of land also rose for a long time, leveled off, and still remains at about the peak. Inputs which have shown the greatest change include fertilizers, machinery, and all manner of productive inputs other than labor and land. Their

[3] Ralph A. Loomis and Glen T. Barton, *Productivity of Agriculture, United States, 1870–1958*, Technical Bulletin 1238, U.S. Department of Agriculture, 1961.

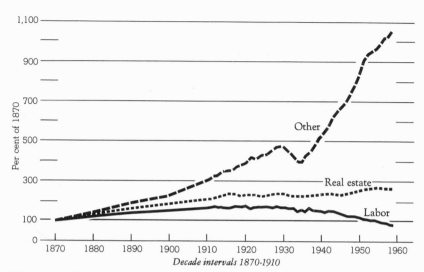

Figure 11. Major input groups in agriculture, from 1870. (Chart from Agricultural Research Service, U.S. Department of Agriculture.)

dramatic rise in recent years is in part an expression of the increased specialization of the productive function carried on by farmers. Whereas agriculture once produced its own power sources, by raising horses and mules, it now buys tractors and trucks; whereas it once produced its own fuel in the form of hay and oats, it now buys gasoline and diesel fuel. Even for such things as seed, which most farmers once produced for themselves, the typical pattern now is purchase from specialized suppliers. Fertilizers, if used at all, were once largely farm manures, at least in some regions. Today, inorganic commercial fertilizers have become the most important source of plant nutrients and are purchased in a volume more than ten times greater than that of 1910. Many such illustrations of the contrasts in the source of productive inputs could be cited.

Another type of specialization is in marketing services. Many farmers once sold and delivered commodities such as dairy and poultry products, fruits and vegetables, and some meats either to consumers or to retail stores. They did whatever grading and packaging was done. Today, marketing services of all kinds, including packaging and processing, are largely carried out by specialized business firms. A complex agricultural business exists which supplies farmers with needed inputs and which processes and markets the farm output.

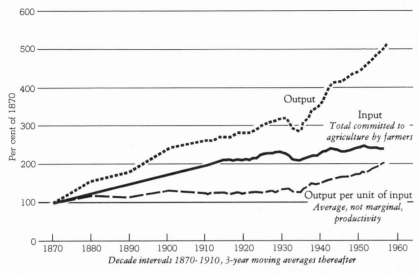

Figure 12. Agricultural productivity, from 1870. (Chart from Agricultural Research Service, U.S. Department of Agriculture.)

The input-output relationship has also changed over the decades. Figure 12 shows how, throughout the period of record, output rose faster than input. However, the difference was not great until about 1935. For the 1870–1911 period, 72 per cent of the increase in farm output could be attributed to increase in the quantity of real estate, labor, and other inputs.[4] Since about 1935, there has been a dramatic upswing in output per unit of input; comparatively little of the total increase in output has resulted from greater inputs. These precise relations depend in considerable part upon just how inputs are defined and measured—no easy task, when the composition of the input mix has changed. But there can be little doubt as to the broad relationships shown in Figure 12.

Of the several factors responsible for the rising output per unit of input, we shall discuss only two: changing technology and improved quality of the human input—and those only briefly.

Technology can be broadly defined as the means of converting inputs into outputs. To use a simple example: At one time, it took 5 to 6 pounds of grain to produce a pound of poultry meat; today good commercial producers of broilers achieve the same result with less than 3 pounds. Or, to take another well-known example—hybrid corn

[4] *Ibid.*

means a greater yield per acre than open-pollinated corn, even when all other factors are the same. In many other instances, a new technological change leads to a different mix of inputs, or to a differently organized farm, or to a different way of production altogether. While technological change can lower production costs per unit of output for the same volume of output per farm as previously, in point of fact the effect of the overwhelming majority of agricultural technological changes has been to increase output. Tractors have been improved, not by making cheaper power available in small sizes, but by making larger and more powerful machines available at reasonable prices. In this case, more power input is combined with a fixed labor input.

While technological change in agriculture has been impressive, even spectacular, the change in the quality of the human input perhaps has been even more important.[5] The farm worker of today is not only far better educated than his father or grandfather, he is probably more alert. Certainly the use of modern machines and agricultural technology would be difficult, if not impossible, with the kind of farm labor that was common a generation or more ago. Just how much of the increased output should be credited to a better working force depends in large part upon how the contribution of the other improved inputs is evaluated, but under any calculations a considerable part of the rising output per unit of input probably must be credited to better workers.

The supply of agricultural commodities is relatively inflexible. Farmers resist strongly any effort to persuade them to reduce their output. As far as the individual farmer is concerned, his output is so small, relative to the total market supply, that his effect upon price is negligible. In deciding how much to produce, he can take the market price as a given factor, and adjust his output to the most profitable level at that price, within the limits or conditions of relevant governmental restrictions or subsidies. Most modern agricultural technology increases output; a farmer can increase his output by using more machinery, new methods, and perhaps more land, without increasing his labor input—which consists mainly of his own labor. Even though he realizes that if many farmers increase their output, prices and the gross value of output are likely to drop, the individual farmer cannot take this into account in planning his output. His reasoning is that he would

[5] Theodore W. Schultz, "Reflections on Agricultural Production, Output and Supply," *Journal of Farm Economics,* Vol. 38, No. 3 (August 1956). See also other articles in other journals written by the same author in recent years.

suffer the reduced price in any event, and it is only through increased volume that he may hold his own or improve his position.

With modern technology, the farmer's ratio of costs to gross income has risen, and this makes him particularly vulnerable to price swings. But voluntary reduction in output is something he can rarely afford as an individual unless he receives government subsidy. Agricultural output rises somewhat more readily in response to favorable price relationships, but even here there is considerable inflexibility. Many other countries of the world have learned that agricultural output cannot readily be adjusted to new growth rates. In this country, a steadily expanding agricultural output, at a rate not very responsive to prices or to government programs, has been, and probably will be, characteristic.

Fluctuations in Agricultural Prices and Income

The demand for agricultural commodities compared to demand for many other commodities does not respond readily to price changes. For all foods at retail, the price elasticity is about −0.30; at the farm, it can scarcely be more than −0.15, perhaps less. This means that a 10 per cent reduction in price at retail will stimulate consumption no more than 3 per cent, and a 10 per cent reduction in the price the farmer receives will increase consumption no more than 1.5 per cent. The elasticity is less at the farm because marketing costs are so high and so relatively fixed that little effect from the decreased price is carried back to the farm. Most people are rather well-fed, and will continue to eat as they are accustomed to eat, even if food prices rise or decline. As a consequence, general price changes have relatively little effect upon their food consumption. For individual commodities, the elasticities are higher, because substitution of one food for another provides opportunities to shift purchases if prices are either higher or lower.

Similarly, the amount spent for food does not change proportionately with income changes. The income elasticity for foods is perhaps as low as +0.20; that is, a 10 per cent rise in income will increase total food consumption no more than 2 per cent. Increases or decreases in food consumption are measured in value, or price terms, for these comparisons; actual poundage of food consumed may not be responsive at all to either prices or income. When incomes rise or prices fall, people eat more of the relatively expensive foods or better qualities of food, not primarily larger quantities.

Inelasticity in demand of these dimensions means that agricultural prices are very sensitive to changes in either supply or basic demand conditions. An unusually abundant crop, one that might lift the index of agricultural output by 5 points or more in a single year, could easily mean a reduction in prices of 30 per cent if there were no price supports of any kind.

One measure of agricultural prosperity is the parity price ratio, which measures the prices farmers receive for the commodities they sell, relative to the prices they pay for the production and consumption items they buy. For comparisons between agriculture and other sectors of the national economy, the parity ratio is similar to the terms-of-trade ratio between different countries. Agricultural economists and others have questioned the propriety of including some of the items contained in this formula, and the use of the 1910–14 or other favorable period as a base period has drawn criticism. However, the year-to-year changes in the parity ratio are informative. The farm parity ratio remained in the 70 to 90 range during most of the later thirties. (See Figure 13.) It rose during World War II and the immediate postwar years, staying at over 100 from 1942 through 1952 and reaching a peak of 115 for 1947. These were years of favorable prices, and also most were years when there were minimum restrictions on output, so that volume of sales was high. After falling below 100 in 1953, the parity ratio declined further and now seems to have stabilized at somewhere around 80.

More or less paralleling these changes in parity ratio, but with additional changes caused by changing general price levels, gross farm income (measured in current dollars) shot up during and immediately after the war. Farm production costs also rose during much of this period, but more slowly, so that net income from farming also rose very sharply from 1940 to 1947. Since 1947, gross farm income has varied from year to year, but with no clear trend; on the other hand, farm production costs have continued to rise, with consequent adverse effect upon net farm income. Farm production costs have risen from 50 per cent of gross farm income in 1947 to 70 per cent today. As farmers have bought more and more inputs, their production expenses relative to their sales have risen. This has made farmers even more vulnerable to price changes, especially for the products they sell.

Total realized net income to agriculture has trended downward since 1947; this downward trend has been almost exactly matched by the downward trend in the number of farms. Realized net income from agriculture per farm has varied somewhat from year to year, with no

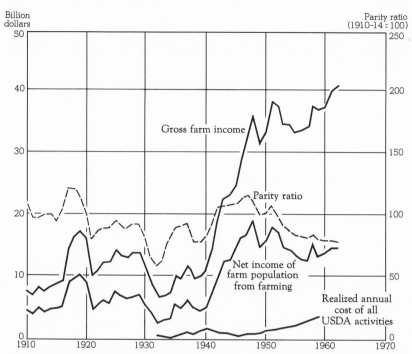

Figure 13. Gross farm income, net income of farm population from farming, parity ratio, and realized annual cost of all activities of the U.S. Department of Agriculture, from 1910.

clear upward trend. The average farmer has just about held his own so far as farm income is concerned, but only by achieving great increases in ouput. His farm is a bit larger, he uses more capital, his managerial skills have improved, and he takes advantage of new technology. This has also been a period in which many farmers have turned to nonfarm employment to supplement their incomes. The period since 1947 has been one of rather steadily rising income per capita among the non-farm population—by almost 70 per cent measured in current dollars. It has been the contrast between rising incomes in other sectors and roughly stationary incomes in the farm sector that has most irked farmers, rather than an actual decline in their incomes.

These comparatively unfavorable price and income relationships have naturally limited the ability of farmers to invest in soil conservation measures, and may well have affected their desire to do so. The interest in such investment, however, is affected by other factors, which are considered in more detail in later chapters.

Rising Land Values and Their Relation to Soil Conservation

Prices of agricultural land have been influenced by several forces since World War II. Chief among these have been: (1) readjustment to the more nearly normal prewar relationship between farm income and land values; (2) extensive investment in land as a hedge against inflation and to reduce federal income taxes; (3) shift of farm land to more expensive nonfarm uses such as suburbs and highways; (4) enlargement of existing farms; and (5) the use of landownership as a factor for participation in federal farm programs.

The very rapid rise in farm income during and immediately after the war was translated into land values slowly and with a considerable lag. In part, this was because the rises in farm income were regarded as temporary, and thus not to be capitalized into land values as though they were permanent and continuing. But, in part, the caution in bidding up land prices during World War II arose because the very serious aftereffects of the massive rises in land prices during World War I were remembered. The speculation in land during World War I and the large debt structure created on the basis of high land prices, left a severe burden of debts to accompany the sharp postwar decline in prices of agricultural commodities.

The rising price level after World War II and widespread fears of uncontrolled inflation led many persons to invest in land. This helped to push land prices up and, in turn, to make this type of investment still more attractive. For those persons in the highest income brackets, an investment need not yield a high annual return to be attractive if it offers the prospect of substantial capital gains in a relatively short period of years. Capital gains, taxed at 25 per cent, leave these persons much better off after taxes than they would be with a larger annual income taxed at 50 per cent. Such persons could afford to speculate in land on the basis of a net interest rate no higher than perhaps 3 per cent if there were a good chance that land prices would rise. Capital gains at this rate might net them as much as other investments at a much higher gross rate of return.

The conversion of rural land to suburbs, highways, or other urban and intensive uses pushed land prices up sharply in the areas directly affected. And it seems probable that price rises in the affected areas tended to spread somewhat to other areas.

Pressures for farm enlargement have been a major force in bidding up prices of farm land. The new technological developments have enabled one man to operate more acres than formerly, and the price-cost squeeze has put pressure on every farmer to increase the volume of his output. For a great many farms, marginal costs of operation were below average costs. An additional 40, 80, or 160 acres of land could be operated more cheaply, acre for acre, than the 80, 160, or more acres in the old farm. As a result, the farmer could afford to pay more for this additional acreage than he could afford to pay for a whole farm, whether his present farm or another one. When a whole farm or some farm land was put up for sale in a neighborhood, either because a farmer retired or for other reasons, the competition among neighboring farmers to obtain this land for addition to their farms often was very keen. Usually prices were bid up considerably in the process.

The farm programs of the Department of Agriculture have been tied to landownership, and have operated to increase land prices. The most extreme form of this is in tobacco, where prices of $1,000 to $2,000 per acre have been paid for tobacco allotments—and these prices are economically rational in view of the income prospects of tobacco production.[6] Acreage allotments, price supports, and other programs have directly affected the price of land used to grow wheat, and perhaps some other crops; they have indirectly affected land for the grain crops such as corn, oats, barley, and sorghum. Marketing agreements for whole milk have raised prices of land used to produce such milk much higher than they otherwise would be.

All of these forces have operated directly to increase the price of that land bought and sold during these years. Actual annual transfers have generally run from 1 per cent to 2 per cent of the total area of farm land; from this some observers have concluded that rising prices of land have not been very important. However, increases in prices paid on land bought and sold spread indirectly to all farm land, and rather promptly. Every farm owner is both a potential buyer and a potential seller of farm land. If land is available at a price which he thinks will enable him to make a profit from buying and operating more land, and if he has capital or credit for land purchase, he will buy additional land. If the price offered for his farm is above his reservation price for that land, he will sell. In a rising land market, reservation prices of farmers who do not sell must rise, at least as fast as the market price rises, or else they sell their land. As a matter of fact, in the rising land market around cities, many farmers have sold

[6] F. H. Maier, *et al.*, *The Sale Value of Flue-Cured Tobacco Allotments*, Technical Bulletin 148, April 1960; and W. L. Gibson, Jr., C. J. Arnold, and F. D. Aigner, *The Marginal Value of Flue-Cured Tobacco Allotments*, Technical Bulletin 156, January 1962; both from Virginia Agricultural Experiment Station, Blacksburg, Va.

and retired or invested their money outside of agriculture, or bought another farm elsewhere. For these farmers, the market price of their land was above their reservation price. Higher land prices, in a genuine market, quickly spread to all land not sold; in this respect the land market is exactly like the stock or grain exchanges. Daily, weekly, or monthly trading in stocks or grain may be only a small part of total supply, but it establishes a price which is effective for the whole supply, not merely that part which is traded. A farmer cannot realize his gain in the price of land until he sells; but, as long as the land market is sufficiently active to establish reliable prices, his gains are real and can be realized whenever he chooses to sell.

Since 1947, the total value of all farm real estate (land and buildings) has advanced by roughly $80 billion in current prices, or has nearly doubled. (See Figure 14.) Part of this rise is related to the rise in the general price level, which has been of the order of 25 per cent since 1947; but most is real, in the sense of a greater rise in agricultural land prices than of prices generally. During this same period, the total realized net income to agriculture has declined and realized net income from agriculture per farm has remained roughly constant. These

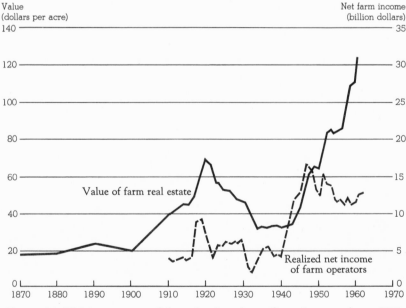

Figure 14. Value per acre of farm land and net farm income, from 1870.

diverse trends between income and value of assets certainly merit attention and concern.

How much of the increase in agricultural land values is caused by each of the five forces enumerated above? As far as we know, no comprehensive studies have attempted to evaluate the importance of these different forces. We hazard the judgment that the effect of the first three forces (readjustment to normal relationships, inflation and tax hedging, and land conversions to other uses) be roughly equal to force 4 (farm enlargement) and to force 5 (government programs). By this, we mean that probably each (force 4 or force 5 or the group of three forces) exerted an influence ranging from 20 per cent to 40 per cent of the total which brought about the rise in land prices. Depending upon the assumptions, as much as half of the rise in land prices could be attributed to the government programs. The situation probably varies considerably among different regions of the United States.

Why should rising land prices cause concern? Is it not advantageous to farmers that the price of their land is rising? The total capitalized value of agriculture in the United States today is nearly $200 billion. At a competitive interest rate of 5 per cent or 6 per cent, this means a total interest charge against land of $10 billion to $12 billion annually. Yet the total realized net income to agriculture is currently only $12 billion to $13 billion. If a full interest return were paid on reported land prices out of current farm earnings, there would be very little left for a payment to labor, whether that of the farmer or hired man. Or, to go at it from the other way, if labor earnings in agriculture were as high as in the average nonagricultural employment, there would be little or nothing left for payment as interest on land values. In this sense, the present structure of prices for farm land is wholly unrealistic. We do not mean to infer that purchasers of farm land who have paid these prices have been economically irrational. Given the rising land market, land purchases were often profitable; or, given the necessity of increasing the volume of business in the farm, purchase of more land may have been the only solution.

But land prices so far out of line with current earnings do raise major policy problems for agriculture and for the rest of the nation. Young men who seek to enter farming must either buy farms, or buy out other heirs on the family farms, at these inflated prices. The difficulty is not that the prices are high in absolute dollar terms; it is that they are abnormally high in relation to current earnings. With land prices jacked up so high, governmental programs for agriculture cannot be changed in a way that would tend to lower land prices, except at great social, economic, and political cost. It is doubtful that any administration could, or would, enter upon a program which would tend to reduce prices of farm land. Farmers are in a difficult position in many instances. They must work hard and manage well merely to earn interest on the market price of their investment, with no net return to their own labor or management skill; or else they get a fair return for the latter at the expense of no current return on the market price of their land.

This unusual capital structure has special meaning for soil conservation. Excessive land prices make some shifts in land use impossible. For example, wheat land cannot be shifted to grass and make a return adequate to pay interest on the present prices; nor can low-grade cropland be shifted to forest, for the same reason. High prices of land are an incentive or pressure on farmers to intensify further, and to increase output more; this will often take the form of more intensive cropping practices than a conservation plan would dictate. High land prices may make it economic to carry out measures to increase productivity of land; it may be cheaper to buy more productivity on present farm acreage than to buy more land productivity in the form of a larger acreage. But the present over-capitalization of land income tends to inhibit further investment in land to maintain productivity at any level, if there is any short-run gain from depleting the soil. One may argue, accurately, that the cost of additional investment in land should be weighed against the additional income from that investment. But when present returns to land are low compared with present land prices, or when adequate returns to land are obtainable only by sacrifice of earnings to labor used in agriculture, then most farmers will be reluctant to invest more for soil conservation ends. The land market seems to take little account of the soil conservation status of the farm; a farm with adequate conservation measures may sell for a little more than one lacking such measures, but the difference is likely to be far less than the cost of installing the measures.

Agricultural Surpluses and Their
Meaning for Soil Conservation

For forty years or more, except for a brief period during wartime, the United States has had an actual, or potential, surplus of agricultural commodities at politically acceptable prices. The actual surpluses have been most apparent for wheat, cotton, corn, and other crops which can be stored for several years. Rigid controls have kept surpluses from appearing for tobacco and whole milk (but have not been so effective for manufacturing milk). Rice has been in actual surplus also, as have some others. One may argue that no surpluses would exist if prices were on a free market basis; declining prices would stimulate consumption or depress production until a new balance would be achieved. Without attempting to estimate the prices at which this might be achieved, or the adjustments it would require, we may simply note that U.S. national policy has not permitted agricultural prices to fall to the levels that would be required, nor does it seem likely that it will do so.

Actual surpluses of some agricultural commodities have attracted considerable attention, in part because they are so obvious. But potential surpluses have been far greater. Imagine what would happen if prices of 110 per cent, or even of 100 per cent, of parity were guaranteed for ten years in advance with no restrictions of output. Total agricultural output could easily double in ten years under such forced draft. Favorable prices would help, but a guarantee of continued favorable prices, taking out all price risks for the guarantee period, would probably be more effective in stimulating output than would higher prices on an annual basis only. It is most improbable that the United States would ever take such steps to stimulate agricultural output, but it is sobering to realize that the potential productivity is so high.

In the early thirties, when the organized soil conservation programs were just getting under way, there were agricultural surpluses, but at the time the surpluses were believed to be temporary, the result of the severe domestic and international economic depression then prevailing. There was a substantial concern over long-run food supply, and over the ability of land to produce enough food to meet future needs. The conservation movement of 1890–1920 and the subsequent soil conservation movement had strong Malthusian overtones—a worry about how man was going to eat. This attitude in the early thirties was the more remarkable in that the consensus of demographers and many others at that time was that the United States was approaching a stationary population at an ultimate level much below the actual population today. Hugh Bennett and other conservation leaders were much concerned about future food supply, but they were also disturbed at the loss of good farm soil by erosion.

The situation today is drastically different, or at least the consensus of the professional community toward it is drastically different. Now, agricultural surpluses dominate the agricultural picture. Many federal agricultural programs are justified, on a cost-benefit basis, because the reduction in costs of storing surpluses will be greater than the increased cost of a proposed program. From this point of view, the surplus is a fact of life which must be coped with, reduced if possible, but scarcely something to be done away with. The consensus among agricultural experts is that farm surpluses will remain with us for twenty years or more; we think it more probable that they shall continue for forty years or more. There have been surpluses over the past thirty years in spite of a growth in population, income per consumer, and foreign trade in agricultural commodities (trade forced and subsidized, but still taking a lot of product) that seemed most unlikely thirty years ago. While some traces of Malthusianism remain among devoted conservationists, generally there has been a very great change in attitude. While some would argue that national efforts at soil conservation are no longer needed, most would not take this position. But few will argue that intensive soil conservation efforts are needed to meet present or imminent demands for agricultural commodities.

The changes that have occurred clearly make it less necessary to preserve every acre of cropland in its present use, or to develop present or potential cropland for highest maximum use, than once seemed the case. Shifts in land use, as part of a conservation program, are now more feasible than they once were. Changes in the number of farms, as well as the changing balance between output and demand for agricultural commodities, point in the same direction. Moreover—though few dedicated conservationists would publicly agree—the immediate national welfare will be less damaged now if some soil erodes, than was expected to be the case. The long-run need for healthy soils and operations geared to maximum agricultural productivity may be as high as ever. Moreover, since the needs for immediate agricultural output are not pressing severely on productive capacity, there is less justification than in the past for exploitative farming, that seeks to obtain present output at the expense of future basic productive capacity. It is today less necessary than in the past for the Soil Conservation Service or any other conservation agency to work with farms and farmers as they find them at any particular moment. Both farms and farmers are changing rapidly, and part of the conservation problem is to keep solutions to the problems in step with the changing times.

PART II

PERFORMANCE
AND EVALUATION

Chapter 6

CLASSIFYING THE LAND
AND ITS EROSION HAZARD

HAVING broadly outlined the background, development, and general approach of the soil conservation work in the United States, our concern now is to fill in the details, to present precise and quantitative information on how the job has been done, what has been done, and what remains to be done. But the detailed story of soil conservation is by no means simple, and to the questions it raises, no brief and unequivocal answer can be given. Instead, it is full of complexities, and the meaning of available information often is not clear. We have examined the evidence, weighed the divergent facts, and drawn the best conclusions we can for inclusion in this and the remaining chapters. Because at almost every turn we have been faced with a lack of basic data on critical issues, we frequently have been forced to stop short of the analyses we would like to make. In such cases, we call attention to the gaps in knowledge, in the hope that better data will be collected in the future.

It is not hard to find excuses, if needed, for failure to document every single aspect of soil conservation, because the activity involves such a multiplicity of related factors. The physical characteristics alone are numerous, diverse, and intricately interwoven when only soil is considered; and they are multiplied rapidly by the addition of land which includes such physical characteristics as geographic location, weather, and vegetative cover. Each of the physical aspects of land can vary over a wide range of values, and the combinations of characteristics is almost infinite—at least so large as hardly to be understood or even imagined by anyone except a soils specialist. A broad regional grouping of land areas according to the most significant or striking features helps to show the enormous variations in basic factors and in their combinations. (See Figure 15.)

Both soils and land not only are affected by geologic history, but they also reflect human history. In the older agricultural areas of the world, as in parts of western Europe, the soil may vary from one side of a fence or property line to the other, resulting from differences in use over generations which have modified the soil. In the United States, irrigation has materially modified many desert soils; with a vastly greater amount of moisture, the microbiological life has completely changed, crop residues and other materials have changed the nitrogen or humus content, and the structure of the soil itself may have been modified. In other areas, changes have also taken place for the better as far as the usefulness of the soil for cultivated crops is concerned. But in the United States, basic soil characteristics often have been modified also in less desirable ways. Unwise cropping may have led to severe erosion, which has carried away a large part of the original soil materials; or land-use practices may have led to increased flooding or to the deposit of soil materials on formerly good cropland; or exploitative harvesting may have so changed the ecological complex of forests and range lands that not only is present productivity lowered, but the original conditions cannot be restored. These various effects of man's actions may be general, evident over rather wide expanses, or localized in particular areas or farms. Some of these changes may be within the reversibility range; others may have left permanent effects so that what originally existed is completely destroyed.

Erosion Hazards in Land Use

Erosion, as we have noted, has received major emphasis from the start of soil conservation efforts—more so in the early years than recently. This is understandable for several reasons; perhaps the most influential of these is the striking visibility of the damages caused by soil erosion. It is not necessary to be a farmer or a specialist to recognize eroded hillsides and muddy streams. But soil erosion, in fact, is considerably more complicated than the public's ready understanding and concern might indicate; it has more causes and more effects—some bad, some good—than first meet the eye.

Among the numerous factors which cause erosion of land, some affect all land within a widespread zone in more or less the same way, although the results locally also depend upon other factors. Primary among the general factors is climate; and, within this category, water and wind are the two most powerful forces. Other localized erosion

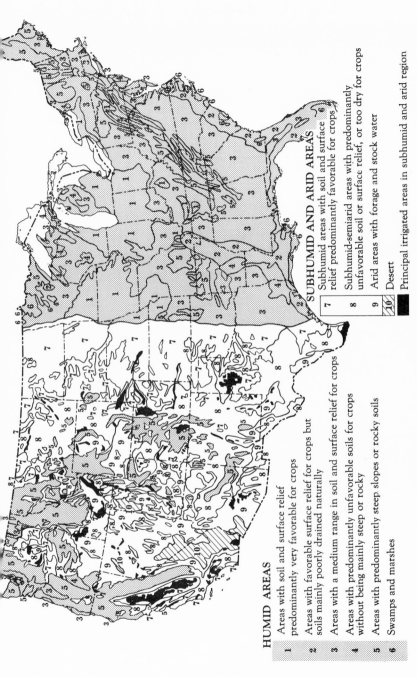

HUMID AREAS

1   Areas with soil and surface relief predominantly very favorable for crops

2   Areas with favorable surface relief for crops but soils mainly poorly drained naturally

3   Areas with a medium range in soil and surface relief for crops

4   Areas with predominantly unfavorable soils for crops without being mainly steep or rocky

5   Areas with predominantly steep slopes or rocky soils

6   Swamps and marshes

SUBHUMID AND ARID AREAS

7   Subhumid areas with soil and surface relief predominantly favorable for crops

8   Subhumid-semiarid areas with predominantly unfavorable soil or surface relief, or too dry for crops

9   Arid areas with forage and stock water

10   Desert

    Principal irrigated areas in subhumid and arid region

Figure 15. Generalized land resource areas. (Map from *Land—The Yearbook of Agriculture, 1958,* p. 17.)

forces or hazards affect specific tracts or spots; these include slope or topography of the land, the specific soil series or soil composition, and the like. Still more hazards arise out of the kinds of land use which are in man's control; his use of land must be adapted to the first two broad classes of erosion hazards.

Rainfall, of course, is a major soil erosion hazard. The striking force of raindrops loosens and moves soil particles; this is proportional to the intensity of the rainstorm, or to the energy contained in it. The quantity of rain which falls not only is related to the intensity of the storm, but also to its duration; if more rain falls, or if it falls faster, than the soil can absorb it, then surface runoff occurs. Moving water carries soil particles, particularly those loosened by the force of the falling raindrops; and the ability of moving water to carry soil material varies as the power of the velocity. That is, water moving twice as rapidly carries four times as much soil material, water moving three times as rapidly carries nine times as much material, and so on. The velocity of the moving water, in turn, is partly affected by its volume, partly by the slope, and also partly by the length of uninterrupted movement. The power of running water to carry soil particles and materials is tremendous. Under extreme conditions, as much as an inch of topsoil can be removed from an entire field in a particularly severe storm, and flash floods can carry great rocks as well as gravel and sand in unbelievable quantities.

Based upon the intensity of storms of different magnitudes, an index of erosion hazard from rainfall is described in an article by W. H. Wischmeier.[1] This index measures the ability of storms to erode soil from unprotected fields. (See Figure 16.) More specifically, Wischmeier says that the erosion index value for a storm "describes the effects of the particular way in which the amount of rainfall energy was combined in that storm with maximum sustained intensity. The numerical value of the index is the product: rainstorm energy in hundreds of foot-tons per acre-inch, times maximum 30-minute intensity in inches per hour. A year's value of the index is the sum of the individual-storm values. The assembled data show that over a long period soil losses from an unprotected field are directly proportional to the values of this rainstorm parameter."[2] Storms vary greatly in intensity and in many locations a large proportion of the soil losses will occur in a

---

[1] See Wischmeier, "Storms and Soil Conservation," *Journal of Soil and Water Conservation,* Vol. 17, No. 2 (March-April 1962); see, also, references cited at end of his article.

[2] *Ibid.,* p. 57.

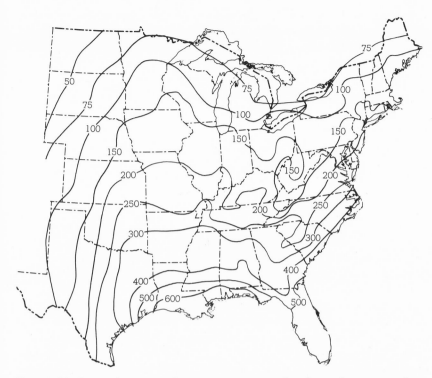

Figure 16. Iso-erodent map showing mean annual values of erosion index for the 37 states east of the Rocky Mountains. (Map from W. H. Wischmeier, "Storms and Soil Conservation," *Journal of Soil and Water Conservation*, Vol. 17, March–April 1962, Figure 1, p. 58.)

relatively few years. In experimental areas, the losses given in the Wischmeier article were as follows: in Oklahoma, 51 per cent of the total loss over 27 years occurred in just 3 years; in Virginia, 81 per cent of the loss was in 3 years out of 17 studied; in Iowa, 40 per cent was in 2 out of 12 years; and in Missouri, 51 per cent was in 3 out of 10 years. Moreover, these years of most severe soil loss were not typically years of extremely high precipitation, but rather of intensive storms.

Figure 16 shows clearly that the hazard of soil erosion from rainfall is much greater in the South than in the Corn Belt; and much more in the latter area than in the Lake States and in New England. In fact, the New England states enjoy the type of gentle rains which characterize the British Isles and much of northern Europe. Many travelers have commented on the green mantle of vegetation which covers the landscape in these overseas countries and the notable lack of soil erosion—comparing this situation with the United States to its disadvantage. It is not necessary to denigrate the beauty of the European landscape and the absence of erosion there, however, to recognize that the problem in much of the United States is vastly different. Northern Wisconsin, for example, has relatively much less erosion hazard from rainfall than in southern Mississippi.

The effect of rain, or the importance of the rainfall erosion index, also depends upon the season in which most of the rain occurs and the normal condition of the land at that season. The South, for example, gets a large proportion of its erosive rainfall during the winter months; and in earlier days, when row crops such as cotton were typically grown in the summer and the land left bare through the winter, soil erosion was often most severe. In more recent times, cover crops have been recommended for wintertime and are now rather widely used; their value is clear from these data on erosion hazards. In the northern Great Plains and in much of the Corn Belt, a large proportion of the erosive rain is in June, when the land is often bare—by then the land has been plowed and planted but the protecting crops are not up, or at least to the point where they can protect the land effectively.

Wind, the other serious climatic cause of soil erosion, usually is a localized or regional hazard. The large region in which it is worse is the Great Plains, where it varies locally. (See Figure 17.) On the average, wind is especially severe in northwest Texas and adjoining parts of Oklahoma, Kansas, and Colorado. (Figure 8 showed this region in which wind erosion damage has been serious.) Another area of severe hazard is in North Dakota, and it is also serious in other localized areas. It is especially damaging to bare soil which has been cultivated with the result that the soil particles are loosened. Such land will often blow very severely when other land in the same general locality will blow much less. Another example of difference is that sandy soils are much more vulnerable to wind than are clays or loams. Since slope, other topographic characteristics, soil type, size of soil particles or structure, and other factors which affect erosion hazard vary greatly within short distances, it is not feasible to show such variations on general maps suitable for a book such as this. However, data on such hazards can be collected and tabulated statistically, and we present some of these data later in this chapter.

The effect of cropping and other land management practices on soil erosion varies widely, depending largely upon other erosion hazards. That is, a particular practice may not be desirable, but it may have

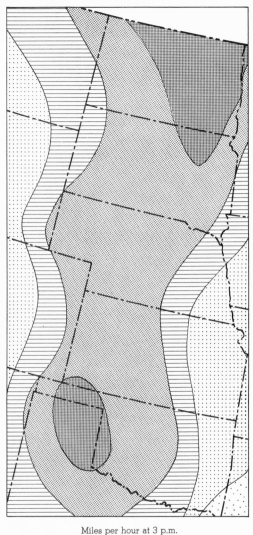

Miles per hour at 3 p.m.

| 16 to 18 | 14 to 16 | 12 to 14 | 10 to 12 | 8 to 10 |

Figure 17. Wind velocity in the Great Plains. (Map from Carter Goodrich, *et al.*, *Migration and Economic Opportunity*, University of Pennsylvania Press, 1936, p. 238.)

little effect upon nearly level land while on steeply sloping land it would be disastrous in terms of resulting erosion; or one practice might be tolerable on loams but wholly unacceptable on sandy loams. Generally speaking, close-growing, sod-forming grasses offer the best protection. Well-forested land may also have good protection from erosion. Clean cultivated row crops—especially if plowed without regard to land slope and without protecting cover during the season after the main crop is harvested—are the worst in terms of soil loss. Various specific practices, such as contour plowing and cultivation, winter cover crops, plowing crop residues into the soil, and leaving trashy fallow in the major wheat-growing areas, may greatly affect the soil losses per acre under any conditions of natural erosion hazard.

Soil and Land Classification for Erosion

Because the factors affecting soil use and erosion may vary so widely and the combinations of factors are so numerous, some grouping or classification of soils is needed. This involves grouping individual units or tracts having very different characteristics into fewer categories, within each of which there are similarities but also differences. Variation within the unit is traded for variation among units. The hope is that the loss in one is far more than offset by the greater ability to understand and grasp the differences among a much smaller number of units.

In soils, the basic unit is the soil series, including its phases and variants.[3] These are determined by careful on-the-spot field studies of depth, soil materials, slopes, and other features. There are several hundred soil series in the United States. Nowadays they often are designated as soil-mapping units, which have been defined as "a portion of the landscape that has similar characteristics and qualities and whose limits are fixed by precise definitions".[4] These, in turn, are grouped into capability units, which are defined as "a grouping of one or more individual soil mapping units having similar potentials and continuing limitations or hazards. The soils in a capability unit are sufficiently uniform to (a) produce similar kinds of cultivated crops and pasture plants with similar management practices, (b) require

[3] Soil Survey Staff, *Soil Survey Manual*, Agriculture Handbook 18, Agricultural Research Administration, U.S. Department of Agriculture, August 1951.
[4] A. A. Klingebiel and P. H. Montgomery, *Land-Capability Classification*, Agriculture Handbook 210, Soil Conservation Service, U.S. Department of Agriculture, September 1961, p. 2.

similar conservation treatment and management under the same kind and condition of vegetative cover, (c) have comparable potential productivity".[5] Of these capability units, those having the same major conservation problem are then grouped into four capability subclasses: erosion and runoff, excess water, root-zone limitations, and climatic limitations. The capability subclasses, in turn, are grouped into eight broad land capability classes. "The capability classes are groups of capability subclasses or capability units that have the same relative degree of hazard or limitation. The risks of soil damage or limitation in use become progressively greater from Class I to Class VIII."[6]

A great deal of the analysis in this chapter and in the several following chapters is in terms of the land-capability classes as used by the Soil Conservation Service (SCS). The basic purpose of this classification is to place soils in broad groups with similar hazards and limitations for use, in light of their tendencies for erosion or in view of their other limitations. Since each broad group includes many different kinds of soil, specific management practices cannot be prescribed for each broad group as a whole, but must be devised for the more specific capability units.

These capability classes are not economic productivity classes, and yet productivity enters indirectly. A generally favorable ratio of output to input is assumed; that is, the general practicability of farming each kind of land is assumed. But specific consideration is not given to economic locations, roads, size of fields, resources of the individual farmer, and so on. The same physical characteristics which put land in a better capability class are also likely to mean that it has a higher economic productivity—for example, deep, well-drained, nearly level soils are likely to have a good rating for each purpose. The land-capability classification assumes a moderately high level of land management—but one within the ability of a majority of farmers and ranchers. When land can be cleared of stones or drained at costs that seem reasonable in light of past experience, it is classed according to its limitations after improvement; but when such improvements cannot be made economically, the land is classed according to its limitations without them. Similarly, lands that have been drained or irrigated in the past are classed according to their limitations after such improvement.

The characteristics of these various land-capability classes are summarized in Table 6. As a general description of land-capability classes

[5] *Ibid.*, p. 2.
[6] *Ibid.*

TABLE 6. Summary of characteristics of land in eight Soil Conservation Service land-capability classes

Item	Class I	Class II	Class III	Class IV	Class V	Class VI	Class VII	Class VIII
Limitations on use for crop production	Few	Some	Severe	Very severe	Impractical	Unsuited	Unsuited	Preclude
Conservation practices required	[a]	Moderate	Special	Very careful	[b]	[b]	[b]	[b]
Practicability of range or pasture improvements	[b]	[b]	[b]	[b]	[b]	Practical	Impractical	Impractical
Permissible limitations singly or in combination:[c]								
Slope[d]	Nearly level	Gentle	Moderate	Moderately steep	Nearly level	Steep	Very steep	Very steep
Susceptibility to erosion	Low	Moderate	High	Severe	Limited	Severe	Severe	Severe
Adverse effects of past erosion	None or slight	Moderate	Severe	Severe	Often slight	Severe	Severe	Severe
Hazard of overflow	Not susceptible	Occasional	Frequent	Frequent	Frequent	Excessive	[b]	[b]
Soil depth	Deep	Less than ideal	Shallow	Shallow	Variable	Shallow	Shallow	Shallow
Soil structure and workability	Good; easy	Unfavorable	Moderate salinity	Severe salinity or sodium	Usually poor	Salinity or sodium	Salts or sodium	Salinity or sodium
Drainage	Good	Correctable	Wetness	Excessive wetness	Poor	Poor	Wet soils	Wet soils
Climatic limitations	None	Slight	Moderate	Moderately adverse	Short growing season	Severe	Unfavorable	Severe
Moisture-holding capacity	Good	Fair	Low	Low	[b]	Low	Low	Low
Stones	None	Few	Few	Few	May be present	Present	Severe limitation	Severe limitation

[a] No conservation practices as such; ordinary management to maintain productivity.

[b] Not relevant.

[c] These are maximum permissible limitations for each class; a particular tract may have no limitation for one factor if other limitations result in its classification in a particular class.

[d] In this context "nearly level" means slopes usually of less than 3 per cent; "gentle" usually means slopes of 1 per cent to 8 per cent; "moderate," slopes of 5 per cent to 16 per cent; "moderately steep," 10 per cent to 30 per cent; "steep," 20 per cent to 65 per cent; and "very steep," over 45 per cent. Classes overlap to include variations within fields or areas and also to allow to some degree for effect of other factors.

SOURCE: Adapted from A. A. Klingebiel and P. H. Montgomery, Land-Capability Classification, Agriculture Handbook 210, Soil Conservation Service, U.S. Department of Agriculture, 1961.

on a national basis, the table necessarily must use generalized terms to meet the extremely diverse land situations which exist in the United states.[7] The same general approach may be used, but with more specificity, on a state or other moderately small region; an example for Kentucky is given in Table 7, which is specific as to latitude, altitude, and annual precipitation. The similarity of the two tables is evident, although the second is specifically related to conditions within Kentucky, and would fit less well in some other state with markedly different conditions—New Mexico, for instance. And this approach could be carried still further, in terms of specific definitions to meet local conditions for a county or other small geographic region.

For any area and by any general definitions, the classification must consider the kind of limitation or limitations and their degree. Some soils will have only one limitation; others will have two or more, but if there are more than one, the separate limitations are less severe. Only rarely, if at all, will a soil have all the limitations listed, and then each limitation would be less severe. The significance of a given limitation will often depend upon the other characteristics of the soil. For instance, a steep slope is more serious for an easily erodible soil than for one more resistant to erosion, or in a region of intense rainfall rather than more gentle rains. It is thus difficult, or impossible, to prescribe exact limits to the boundaries of the various groups for application on a national scale; these have to be worked out specifically in each locality.

The best land is in capability Class I, which contains the nearly ideal soils of the United States. There are no significant limitations for cropland use, beyond those farming measures that should normally be undertaken to maintain productivity. This land is nearly level, has low susceptibility to erosion, is not subject to flood overflows, and has deep soils, well-drained with good texture and workability. These are the nearly ideal soils of the United States.

Class II soils have some, but not severe, limitations for crop production; they require moderate conservation practices. They may have gentle slopes, or moderate susceptibility to erosion, or show moderate effects of past erosion, or be subject to a moderate hazard of occasional overflow, or have less than ideal soil depth, or have somewhat unfavorable soil structure and workability, or have some undesirable but correctable wetness, or have some modest climatic limitation—or there

[7] Personal communication from A. A. Klingebiel, January 21, 1964, states that there are "over 100,000 different kinds of soil" which can be grouped into eight land-capability classes or into twenty-five different land-capability subclasses.

Performance and Evaluation

TABLE 7. Principal characteristics of the land-treatment units in each land-capabili♦

Land-capability class	Land-treatment unit	Effective depth	Texture of surface soil
Suited for cultivation: I Very good land; few or no limitations; can be cultivated safely with ordinary good farming methods	Nearly level, well-drained bottom land	Deep	Medium
	Nearly level, well-drained upland	Deep	Medium
II Good land; moderate limitations or hazards; can be cultivated safely with moderately intensive treatments	Gently sloping soils on limestone	Deep	Medium
	Gently sloping soils on shale	Moderately deep	Medium
	Nearly level, or gently sloping imperfectly drained soils	Deep	Medium
	Nearly level, or gently sloping imperfectly drained bottom land	Deep	Medium
III Moderately good land; severe limitations or hazards; can be cultivated safely with intensive treatments	Moderately sloping soils on limestone	Deep	Medium
	Nearly level to moderately sloping shallow soils with plastic clay subsoils on limestone	Shallow*	Heavy
	Moderately or strongly sloping soils on shale	Shallow or very shallow	Medium
	Nearly level, gray wet soils	Deep	Medium
IV Fairly good land; very severe limitations or hazards; suited for some forms of limited cultivation	Steep or stony soils on limestone	Shallow or moderately deep	Medium
	Steep, shallow soils on shale	Shallow	Medium
	Soils with heavy plastic subsoils, nearly level, on limestone	Moderately deep	Medium
	Wet soils on bottom lands	Deep	Medium
Generally not suited for cultivation, but suited for grazing or forestry: V Few natural limitations for growth and utilization of permanent vegetation	Wet bottom lands; not suited for cultivation	Deep	Medium
VI Growth or utilization of vegetation moderately limited by land characteristics; generally good or fairly good grazing land or woodland	Very good steep, eroded, or stony land on limestone	Shallow to deep	Moderately light
	Bottom land or low land subject to extreme deposition or erosion	Deep	Medium
VII Growth or utilization of vegetation severely limited by land characteristics; generally fair or poor grazing land or woodland	Stony, rough, or severely eroded land on limestone	Shallow	Medium
	Stony, rough, or severely eroded land on shale	Very shallow	Moderately light
Not suited for cultivation, grazing or forestry: VIII Suited for wildlife, recreation, or watershed protection	Rock outcrop; stream banks, ditches, cuts, fills, etc.		

ᵃ Latitude is about 39.5° N; altitude, 500 to 900 feet; and precipitation, about 37 inches. The lim iting characteristics most significant in determining the land-capability class are indicated ▶ one star; the distinguishing characteristics that differentiate land-treatment units within a lan▶

ass in a Kentucky soil conservation district[a]

Permeability		Native fertility	Available moisture capacity	Slope	Degree of soil erosion	Excessive wetness	Overflow hazard
Subsoil	Substratum						
Moderately rapid	Moderately rapid	High	High	Nearly level	None	None	Occasional**
Moderately rapid	Moderately rapid	High	High	Nearly level	Slight	None	None**
Moderately rapid	Moderately rapid	High**	High	Gentle*	Slight	None	None
Moderately rapid	Moderately rapid	Moderate**	Moderate**	Gentle*	Slight	None	None
Moderately low*	Moderately slow*	Low**	High	Nearly level or gentle	Slight	Moderate*	None
Moderate*	Moderate*	High**	High	Nearly level	None	Moderate*	Occasional**
Moderately rapid	Moderately rapid	High**	High**	Moderate*	Slight or moderate	None	None
Moderately low		Moderate	Moderate	Nearly level to moderate	Slight or moderate	None	None
Moderately rapid	Moderately rapid	Moderate**	Moderate**	Moderate*	Moderate	None	None
slow*	Very slow*	Low	Moderate	Nearly level	Slight	Very wet*	None
Moderately rapid	Moderately rapid	High**	Moderate**	Steep*	Moderate or severe	None	None
Moderately rapid	Rapid	Moderate**	Low**	Steep*	Moderate or severe	None	None
Slow*	Slow*	Moderate	Moderate	Nearly level	None	Moderate*	None
Moderately low	Moderately slow	Moderate	High	Nearly level	None	Very wet*	Frequent*
Moderately slow	Slow	High	High	Level	None	Very wet, not drainable*	Very frequent
Moderately rapid	Moderately rapid	High	Moderate	Very steep*	Moderate or severe	None ·	None
Variable	Variable	Moderate	Moderate	Level or undulating	Severe erosion or deposition**	Moderate	Very frequent
Rapid	Rapid	Moderate**	Low	Very steep or rough*	Moderate to severe	None	None
Rapid	Very rapid	Low**	Low	Very steep or rough*	Moderate to severe	None	None

apability class are indicated by two stars.
SOURCE: Based on a table provided by the Soil Conservation Service, U.S. Department of griculture.

may be some combination of these factors. The key element in each is its moderation. For example, if Hanford fine sandy loam in California has a 2 per cent slope, or less, it is in capability Class I. The same soil with a slope from 2 per cent to 6 per cent, or with some erosion occurring, would be placed in Class II. Also in California, Chino silt loam with no salt or alkali is Class I, but with slight salinity or alkali it becomes Class II. Holdrege silt loam in Nebraska with a slope of 2 per cent or less and an "erosion class" of 1 or 2 is land-capability Class I when irrigated; with the same, or slightly worse, characteristics if unirrigated, it is Class II. Memphis silt loam in Tennessee with a slope of 2 per cent and no more than erosion class 1 is capability Class I; with the same slope and erosion class 2, or with 2 per cent to 5 per cent slope and erosion class 1, it is capability Class II. Honeoye silt loam in New York with 3 per cent or less slope and erosion class 1 or 2 is capability Class I, but, with 3 per cent to 8 per cent slope and equal erosion, is Class II. Many other illustrations could be cited, of course. The nature of the required conservation practices is often simple—plowing and cultivating on the contour, or turning under certain crop residues, vegetated water-disposal areas, and so on—or may be somewhat more involved, as terraces or strip cropping. By any standards, these are good croplands, generally productive and easily kept in permanent farming.

Class III soils have more severe limitations for permanent crop production, and they often require special conservation practices. Soils may be moderately steep; or may have a high susceptibility to erosion, or show severe effects of past erosion; or have frequent overflow; or be shallow above bedrock or hardpan, or have some other limitation to the rooting zone and water storage; or have moderate salinity; or have restricting wetness even after drainage; or have moderate climatic limitation; or have low water-holding capacity. And there may be some combination of these limitations, but with each specific limitation in lesser degree. The difference between capability Class II and Class III is a matter of degree; this takes different specific expression in different localities or situations. Tama silt loam in Illinois with 2 per cent to 7 per cent slope, but only erosion class 1 to 2, would be in Class II; the same slope with class 3 erosion, or slopes of 7 per cent to 12 per cent, but with class 1 or 2 erosion, would be capability Class III. A similar situation exists for Memphis silt loams in Tennessee, or Honeoye silt loams in New York, or Williams loam and clay loam in North Dakota, or Hanford fine sandy loam in California. While these are not the best soils for crops, they can be used permanently for crop production if adequate conservation practices are followed. The practices required on Class III land are more complex or costly, or both, to install and to maintain than on Class II land. Terraces, grassed waterways, special crop rotations including more use of close-growing crops, better use of crop residues, and other practices are necessary. Such lands cannot be used as intensively for clean-tilled row crops, and still provide adequate conservation, as either Class II or Class I land.

Capability Class IV lands have very severe limitations for crop production, and if so used must have very careful management to preserve their productivity and to avoid serious loss. They may be relatively steep, have a severe erosion hazard, show the effects of severe past erosion, or be subject to frequent overflow; they may have shallow soils, or severe salinity or sodium conditions, or excessive wetness with a continuing hazard of waterlogging even after drainage, or moderately adverse climatic conditions, or low moisture-holding capacity. Again, if they have some combination of these limitations, each will be in lesser degree. Broadly speaking, this class of soils is marginal for crop production. Some may be farmed continuously, but can be preserved only with the most careful management practices; others may be cropped occasionally, being kept in tight pasture cover or otherwise protected except during the crop years; and still others may have such serious limitations that farming them ordinarily would not be very profitable. Some are nearly level and not subject to erosion, but so poorly drained that they have low crop productivity. Others are steep enough to present a serious erosion hazard. Some of these steep lands may be well adapted to produce tree fruit if adequately protected against erosion. In subhumid and semiarid areas, soils in this class may produce a good yield in years of above average rainfall, a low yield during years of average rainfall, and failures during years of low rainfall; but the soil must be protected against blowing in all years, if severe wind erosion is to be avoided. To illustrate Class IV, Tama silt loams in Illinois, Honeoye silt loams in New York, or Holdrege silt loams in Nebraska may be included with slopes up to 12 per cent, 25 per cent, and 10 per cent, respectively, if past erosion has not been too severe. Kalkaska sand in Michigan and Wisconsin and Lakeland sand in South Carolina, each of low moisture-holding capacity and low fertility, are so classed. Chino silt loams in California, with strong salinity or alkali or a water table that is seasonally high, are also in this class. Thus, many different kinds of soils, each with its own particular limitations and hence its need for particular management practices, is included in Class IV.

Class V soils are not practical for production of crops. These soils are typically subject to such frequent and severe overflow that crops cannot be grown successfully. In the case of some, the growing season is too short; others are too rocky for cropping. While these lands must be downgraded for cropping, they do not present a serious erosion hazard.

Class VI and Class VII soils also are not suited for cropping. They may range up to steep or very steep, or may have severe erosion hazard, or show severe effects of past erosion, or be subject to frequent and serious overflow, or be shallow, or have salts or sodium accumulations, or have serious climatic limitations, or be rocky, or have some combination of these deficiencies—in which case, the several kinds of deficiencies are each less serious than if taken singly. Soils in Class VI may sometimes be used for special types of crop production; for such uses as orchards if tightly sodded or for blueberries. The distinguishing difference between soils in Class VI and Class VII is that the former can be improved practically by use of range or pasture improvements such as seeding, liming, fertilizing, water control, and the like, while the latter cannot. The differences between these two classes of land for pastures or ranges or forests are thus matters of degree, roughly corresponding to the difference in degree between Class II and Class III for crop production.

Of all soils and forms of land, those in Class VIII are the least productive and the most serious in erosion hazard. This class includes badlands, rock outcrops, sandy beaches, river wash, mine tailings, and other nearly barren lands. Not only can these soils not be cropped successfully, they are uneconomic for grazing and forestry. They generally produce none, or little, grass or other forage, or trees. Some Class VIII lands have scenic values, but many present a serious erosion hazard, not only to the land so classed but to downstream or downwind areas. These are the "wastes" of the United States. Fortunately, there is a very small area of such land within the forty-eight contiguous states; however, there is more in Alaska than in all the other states combined.

As a means of dealing more explicitly with the great variations within each land-capability class, the grouping into capability subclasses is very useful. As noted earlier, there are four of these: The subclass including erosion and runoff is indicated by e; that for excess water by w; for root-zone soil limitations by s; and climate, by c. Subclass *e* is made up of soils where susceptibility to erosion is the dominant problem or hazard; where two or more limitations are evident, but of approximately equal importance, the *e* limitation takes precedence over the others. Subclass *w* is made up of soils where excess water is the dominant limitation or hazard; poor soil drainage, wetness, high water table, and overflow are the criteria for determining which soils belong in this subclass. Where two or more limitations are approximately equal in importance, *w* is subordinate to *e* but takes precedence over *s* and *c*. Subclass *s* has serious soil limitations, such as shallowness, stones, low moisture-holding capacity, low fertility that is difficult to correct, or salinity or sodium, or some combination of these, within the rooting zone of plants. Where two or more kinds of limitations are more or less equal in importance, *s* is subordinate to *e* and *w*. Subclass *c* is made up of soils where the climate (temperature or lack of moisture) is the only major hazard or limitation. While these subclasses still contain many different kinds of soil situations within each, the kinds of problems in the broad capability classes are largely differentiated by these subclasses.

In terms of our definition of resources, the land-capability classification of SCS might appear to be simply a classification of the "quality or characteristic of nature." But it is necessarily more than that. Presently known and used agricultural techniques are included, consciously or unconsciously; economic feasibility enters, in a general way, since soil classification is based upon the experience of farmers with each soil or with similar soils. The general goal of maximum income, or at least of income up to some acceptable standard, is unconsciously assumed. We find it impossible to imagine a classification of soil characteristics which does not consider such factors to some extent. So far as possible, the SCS classification is based upon physical characteristics; yet the characteristics and the magnitude of the various variables have been chosen, as they must be, with consideration of such factors as their technological and economic relevance.

The SCS land-capability classification system was worked out by soil specialists particularly concerned with the conservation problem. The results of research on soil erosion and other problems were utilized in drawing up the specifications of the different land classes. For a great many parts of this classification system, judgment is required as to the importance of particular factors, and for applying this classification system to any locality, as precise standards as possible must be drawn up. For example, the hazard arising out of steep slopes depends in part upon the kind of soil and in part upon rainfall intensities; but it is necessary to specify just what slope, just what soil texture, just what past erosion effect, and so on. In drawing up these precise standards, full use is made of experimental results and of observed experience elsewhere. But sometimes classifiers may not have the same judgment of the importance of different degrees of the various hazards.

It seems probable that the capability classification system as used by the SCS represents a point of view which is at least oriented to conservation. That is, the men responsible for it are deeply concerned over the soil conservation problem, and it would seem likely that they would weigh more heavily the various risks than would others whose concern is not so direct. An effort has been made to obtain objectivity in the soil classification, and we have no evidence to suggest that deliberate bias entered. But we can, perhaps, assume that the hazards to soil conservation revealed by this land classification have not been underrated, and unconsciously may have been slightly overrated.

The Conservation Needs Inventory and Its Data on Land Capability

Since the Soil Conservation Service developed the concept of land-capability classification during the thirties, it has applied the classification to numerous areas of the United States. But it was not until 1957 that the Department of Agriculture embarked upon the National Inventory of Soil and Water Conservation Needs.[8] This Conservation Needs Inventory (CNI) was a Department-wide undertaking, under a committee representing several agencies within the Department; however, it was of chief interest to the SCS and was largely carried out by that agency. The whole undertaking assembled "data about four major aspects of the Nation's agricultural land, as follows: (1) *Land capability,* by class and subclass; (2) *land use,* actual in 1958 and expected in 1975; (3) *conservation treatment needed* on land in each land use; and (4) *small watershed projects needed.*"[9]

The CNI is the most careful attempt yet made in the United States to obtain quantitative and comparable national information about land and its conservation problems, and at the same time it is also dis-

[8] *Policy and Procedure for Development of National Inventory of Soil and Water Conservation Needs,* U.S. Department of Agriculture, August 1957.

[9] Conservation Needs Inventory Committee, *Basic Statistics of the National Inventory of Soil and Water Conservation Needs,* Statistical Bulletin 317, U.S. Department of Agriculture, August 1962 (italics in original).

appointing. Carefully using tested professional techniques, random sample areas were chosen in each county with the help of well-qualified university statistical laboratories. The standard size of a sample area was 100 acres in the thirteen northeastern states (generally where the rectangular cadastral survey does not apply) and 160 acres elsewhere (or a standard quarter section in the rectangular cadastral survey states). The basic sampling rate was 2 per cent, which provides an acceptable degree of reliability in a county of one-quarter to a half million acres. Because of the way the samples were selected, the sampling error was known in advance. An item reported for 1 per cent of the area has a 5.1 per cent sampling error for a state, and an 0.9 per cent sampling error nationally; an item reported for 5 per cent of the area has a sampling error of 2.3 per cent for a state and 0.4 per cent nationally; and progressively smaller sampling errors as items are for larger areas. Data obtained from sample areas can be expanded with reasonable confidence to obtain totals for counties, states, and the nation, and this was done with the Conservation Needs Inventory.

Inventories of the sample areas were made in the field, using techniques of survey and classification developed over the years. The information so obtained was supposed to be tabulated according to an established procedure, including the use of modern statistical processing machinery. Full use was made in the field of earlier soil surveys, of aerial photographs, and other information. Inventories of all land, except the larger federally owned areas and certain urban and built-up areas, were made in this way. Our concern in this book is primarily with the contiguous forty-eight states, omitting both Alaska and Hawaii as well as Puerto Rico and other possessions; for these states, slightly more than 75 per cent of their total area was covered.

The deficiencies in the CNI were primarily in its execution and in the subsequent analysis of the data which had been collected. No additional funds were available to the Department of Agriculture or to the SCS for the Inventory; and no specific provision was made for reduction in the other duties of many of the persons connected with it. To a large extent, it represented an added load of work. Many of the persons involved in it, especially at the field level, drew the not unreasonable conclusion that it was a job of low priority within their agency. The work lagged in some areas, and in others seemed to have been done to less careful standards than were planned. Some states did not forward their data to the central statistical laboratories for processing, but prepared their own reports. Differences between counties or soil conservation districts and between states were to be expected, but it seems that they were not fully eliminated in the final results. Perhaps most important of all, the full potentialities of the CNI data have never been realized because of inadequate statistical analysis.

At several points in this chapter and in following chapters, we raise important questions of policy that might have been answered from the CNI data but have not been because of inadequate statistical analysis. In spite of these disappointing deficiencies, however, these data are the best available, and in fact are quite helpful in answering many important questions.

The three-quarters of the total acreage in the forty-eight contiguous states included in the CNI are grouped by land-capability classes in Table 8. It is safe to assume that most of the uninventoried acreage is

TABLE 8. Area of land in Conservation Needs Inventory according to land-capability classification   (Millions of acres in 48 contiguous states)

| Land classification[a] | Total inventoried area | Area in subclasses, according to dominant limitation[a] | | | |
		Sub-class e (erosion)	Sub-class w (excess water)	Sub-class s (unfa-vorable soil)	Sub-class c (climate)
Class I[b]	36.2				
Class II	290.1	149.8	86.4	33.1	20.8
Class III	310.8	178.0	73.7	48.2	10.9
Subtotal, best arable classes	637.1	327.8	160.1	81.3	31.7
Class IV	168.7	95.9	27.5	41.1	4.2
Subtotal, arable classes[c]	805.8	423.7	187.6	122.4	35.9
Class V	43.0	.5	38.7	2.0	1.8
Class VI	276.8	146.8	7.1	97.9	25.0
Class VII	294.2	162.0	6.8	114.2	11.2
Class VIII	26.7	4.8	5.2	15.8	.9
Subtotal, poor classes	640.7	314.1	57.8	229.9	38.9
Total classified area[d]	1,446.5	737.8	245.4	352.3	75.8

[a] See Table 6 and discussion in text for meaning of these land classes and subclasses.
[b] Land in Class I has no significant limitations.
[c] Some land in poorer classes than these can be used for production of special crops, such as orchards with complete grass sod cover, cranberry bogs, and so on.
[d] Totals exclude 1.4 million acres inventoried but not classified, and excludes federally owned land not in crops (396 million acres), most of which would fall in Classes VI and VII, and excludes also nearly 51 million acres of urban and built-up land and nearly 7 million acres of water area in small streams and lakes which are often included in land area—or a total of 455.4 million acres. The totals for the subclasses do not equal the total given for inventoried area in the first column because Class I acreage is not broken into subclasses.
SOURCE: From *Basic Statistics of the National Inventory of Soil and Water Conservation Needs,* Statistical Bulletin 317, U.S. Department of Agriculture, August 1962, p. 101.

in Classes VI, VII, and VIII. There is very little cropland, actual or potential, in the larger areas of federally owned land that were excluded from the Inventory. Some land capable of growing crops is included in the urban and built-up areas, but this has been effectively taken out of agricultural, grazing, and forestry use for an indefinite period.

Of the classified area, only 2.5 per cent is in Class I, with no limitations for crop production; but 44 per cent is in the three best arable classes. Over half of the acreage in these three classes has an erosion hazard, nearly equally divided between that with a slight hazard or slight present erosion and that with a more moderate potential or past erosion. The next most common hazard for these three classes is excess water or wetness, which affects about one-fourth of such lands, again about equally divided between slight and moderate degrees of wetness. Unfavorable soils and climatic hazards follow in this order. One fact stands out in these data; the relative acreages follow in the same sequence as the priority order among the limitations. If a different order of limitations had been chosen, how different would the figures have been? Thus, subclass *e* contains not only the land on which erosion is the sole or dominant limitation, but also the land on which this limitation is approximately equal with one of the others. Had *w* or *s* been considered dominant, how much would subclass *e* have been reduced?

In terms of our discussion in Chapter 1, there is a definite problem of maintaining productive capacity on the lands of subclass *e*. Unless proper practices are followed, erosion of varying degrees will take away some of the soil materials. In some cases, at least, this will seriously and irrevocably damage the productive capacity of the land, while in other cases it will have less serious effects. The problem in the other subclasses is less one of maintaining original productive capacity than it is either to build new capacity into such lands (by protection from floods, drainage, removal of stones, or in other ways) or to manage the lands so as to minimize the unfavorable economic effects of these limitations.

In addition to that in the three best classes, about 12 per cent of the total inventoried acreage is in Class IV, the marginal croplands. These exhibit similar, but more severe, limitations. While erosion is the most important limitation here, unfavorable soil comes second, and excess water comes third.

It would be especially interesting to know where the soils of these different classes are to be found in the United States, but this is among the analyses of these data which have not yet been made. In the absence of such analysis, it is necessary to turn to maps showing the distribution of such lands when actually used for crop production. The

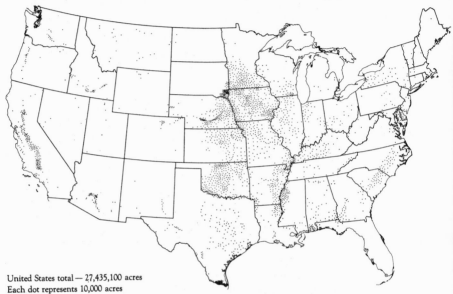

United States total — 27,435,100 acres
Each dot represents 10,000 acres

Figure 18. Land-capability Class I, 1958 cropland acreage.

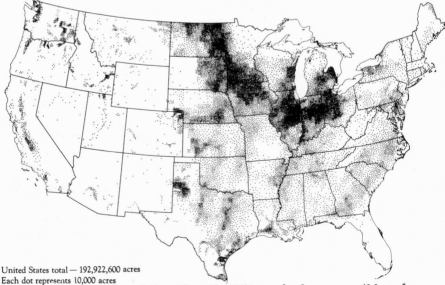

United States total — 192,922,600 acres
Each dot represents 10,000 acres

Figure 19. Land-capability Class II, 1958 cropland acreage. (Maps from Economic Research Service, U.S. Department of Agriculture.)

acreage in each of the first four classes, which was also in crops in 1958, is shown in Figure 18 through to Figure 21. In Chapter 7, we consider how much of each class is in crops, but the distribution of the

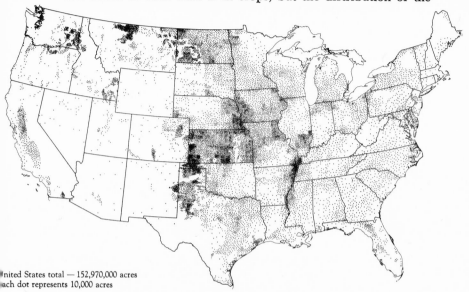

United States total — 152,970,000 acres
Each dot represents 10,000 acres

Figure 20. Land-capability Class III, 1958 cropland acreage.

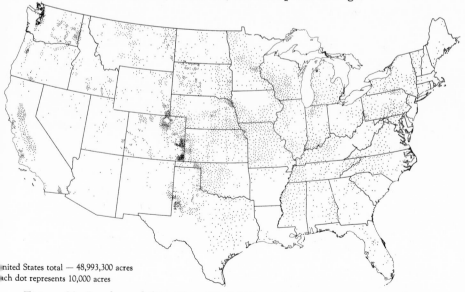

United States total — 48,993,300 acres
Each dot represents 10,000 acres

Figure 21. Land-capability Class IV, 1958 cropland acreage. (Maps from Economic Research Service, U.S. Department of Agriculture.)

crop acreage may give some indication of the distribution of all land in the class.

Cropland in Class I is found in many states—in the coastal plain of the Southeast, in the Mississippi delta, through central Oklahoma, in the western part of the Corn Belt, in the Central Valley of California, and elsewhere in smaller tracts. Although Class II cropland is found throughout the nation, the largest areas are across the Corn Belt and up into eastern North Dakota. Class III cropland is also widely distributed, but relatively more is found in the Great Plains. Some very interesting contrasts along state and county lines show up between Class II and Class III. Note how much Class II land is found across southern Nebraska and how little in northern Kansas; or how little Class III land there is across southern Nebraska and how much across northern Kansas. Another contrast is the bunching of one or the other of these two classes in small areas—counties or soil conservation districts—especially within the Great Plains, but to some extent elsewhere. Such sharp differences make one wonder whether these differences reflect the classifier as much as the soils classified.

Class IV croplands are also scattered across the country, somewhat more in eastern Colorado than elsewhere, and proportionately more throughout the Great Plains. However, only a little more than a fourth of lands in Class IV are in crops, so it is at least doubtful that the distribution of all lands in this class follows the pattern shown by its lands in crops.

This leaves about 44 per cent of the classified area as nonarable; if all unclassified lands are also assumed to be nonarable, then the total nonarable area would be about 54 per cent of the total area of the forty-eight contiguous states. The term "nonarable" for this estimate is used in both an economic and a physical classification. That is, in less well-endowed countries of the world, some of the lands classified in the United States as nonarable would be farmed. For example, the virgin lands of Siberia which the Soviet Union has plowed up in the past decade may be no better than some of the nonarable lands of the Great Plains. Or some of the lands subject to a relatively large amount of overflow might be farmed if the pressure on land were severe enough. No maps are available to show the location of all land in Class V through Class VIII. Figure 22 in Chapter 7 shows cropland in these classes, but it indicates the location of only a minute fraction of such lands since they are so unsuited for the production of most crops.

Although there are major differences in the generosity of cropland endowment between different parts of the nation, yet these data and maps make it clear that no one area, or even a few areas, have either all the good land or all the poor land. Land of different classes is typically intermingled in the same state, usually in the same county, and often on the same farm. In one sense, this greatly complicates the problems of soil conservation—they do not sort out simply and cleanly along boundary lines of governmental or landownership units. But it also greatly simplifies the problems in other ways—adjustments of land use to land capabilities is not a problem of some regions or some farms only, but arises widely; and some better lands are to be found nearly everywhere, offering opportunities for use as croplands.

When the native plants associated with different soil conditions in grazing and forested areas are considered, the situation becomes more complex than ever. Forest and range lands, like the croplands, are subject to the manifold variation of the soils; in addition, the kind of plant cover involved can also be diversified and subject to a great many influences, particularly climatic ones. Typically, many different kinds of plants are found in a general soil and land situation; each kind has its own particular requirements for growth. Some are highly important for human use; others, less so or unimportant. Within each particular species, there may be great variation among individual plants, as to age, size, vigor, annual growth rate, and other factors. The many different individuals and different kinds of plants make up a community and, together with the animals, an ecosystem. Relationships among individual plants of the same or different kinds may take many forms—competitive, complementary, symbiotic, parasitic, and others.

The nature of the plant community not only depends heavily upon soils and climate in all their manifold variations, but it also depends upon man in varying degree. Man, whether primitive or modern, can be considered as part of the natural background of the plant community or as a disturbing force. Modern man certainly has capacities to modify the plant community which his forebears lacked. But the latter often exerted more influence than is sometimes realized. Fire, whether caused by lightning or started deliberately or accidentally by man, has been a vital force affecting vegetative cover in many regions. It has been least important in the desert areas where fuel supply was often insufficient to carry a fire, and in very high rainfall forests where everything was too green and wet to burn readily. In much of the United States, however, forest and range fires can and do occur, with major effects upon plant cover.

In the short run—meaning, for this purpose, a few decades—present vegetative cover is a basic resource which may limit the use of forest and range lands. Over a longer period, man can change the kinds and amounts of vegetation, within some limits and usually at some cost. If the aim is to harvest trees 100 years old, then this will obviously take 60 years if the oldest trees at present are but 40 years old. If a certain type of grass is needed for grazing, it probably can be established or re-established, if the climate and soils are suitable, but this will take time and usually money. Grazing areas of the United States typically lack annual precipitation, and respond slowly to most measures aimed at their improvement. Tree growth also is often slow, and changes in forests take much time.

It would be possible, but not fruitful, to consider the kind of vegetative cover existing in the United States when the white man first came. Knowledge of conditions then lacks precision, but in any case the changes have been so profound that present vegetative cover is vastly different. Not only have some forests been cleared for farming and some grasslands plowed up, but changes in composition and density of cover have been very great. In Chapter 8, we consider the present condition of forests and grazing areas, after the discussion in Chapter 7 about the use of land.

Use of the Land-Capability Classification for Urban Development Planning

Although the Soil Conservation Service developed the land-capability classification for use in agricultural areas where farming is the dominant land use, this classification also has major usefulness for development planning in suburban areas. Many of the same basic factors about soil are involved in each case, and the same kinds of information are usable, although often in rather different combinations.

The internal drainage of the soil is a critical factor in the use of septic tanks in suburban areas. If an attempt is made to use such tanks in heavy clay soils, there may be serious backflow in periods of heavy rainfall; or the soils may become polluted because of inadequate capacity to deal with effluent; or other problems may arise. In some sandy soils, effluent may pollute the sources of water supply if reliance is placed upon wells for domestic water. General overflow problems, such as a land-capability classification can reveal, are also serious in some suburban areas. The ability of different soils to bear stresses can be a major factor for some types of industrial and other urban land uses, although it is generally less important for residential use.

The slope of land and erosion hazards may be important for many kinds of urban land uses. While slope is obvious to any engineer or builder, erosion hazard is less so. In many areas, the amount of sediment entering streams during the construction period is very high, especially if the natural erosion hazard on the land is high. If a suburban area is to be used for residential purposes, then data on soil depth, soil fertility, and soil texture may be useful to the planners. The intensity of land use, or the amount of inputs per acre, is often much higher in residential areas than if the same land were used for crops, even when allowance is made for the fact that a considerable part of the land is covered with impervious streets, rooftops, and driveways in the suburban districts.

The Soil Conservation Service in recent years has developed direct working relations with urban planners and builders, and is making the information on land-capability classifications available to and usable by these urban interests. In some cases, soil conservation districts have also developed working relationships with such groups. This reflects an incidental, but important, use of information originally developed for use in rural areas only.

Chapter 7

LAND USE IN RELATION
TO CAPABILITY

How WELL does the actual use of land in the United States conform to the land-capability classes? Are farmers trying to grow crops on some, or much, unsuitable land? Are there good lands capable of continued crop production but not used for this purpose?

For nearly a hundred years, the Census of Agriculture, and the annual reports of the U.S. Department of Agriculture, in particular, have presented statistics on the area of land used for various purposes. Except in a very general way, these data on land use have not been matched to information on land capability. Numerous studies of local areas have considered land characteristics or quality and the use of land; but these have been confined to relatively small areas, without general or national coverage, and most of the various studies have used different methods so that their results are not readily comparable.

Generally speaking, farmers have gradually concentrated their crop operations on the better lands in the United States. Throughout U.S. land history, men have always sought to grow crops on the "best" land available to them. At any given time, "best" involves many features— not only soils and climate, but also transportation and markets, farm technology, and even the opportunity to acquire land in various localities. Many mistakes have been made in choosing land, of course; it has been a process of trial and error. A great deal of land in the United States has been planted to crops at some time and then abandoned—some to be cropped and abandoned again. New technology frequently has made land usable and valuable which previously could not be used at all, or only with difficulty; and, through competition for the markets, this process often has made other lands much less valuable than previously.

The process of adjustment has been continuous. In shifting to the use of better lands, it has been necessary to overcome much resistance to change and many obstacles. Farmers have investments in the form of buildings on poor land sometimes, and good economics may dictate employment of these investments as long as they are usable. But the greatest barriers to change have been human: farmers have not realized that better land existed elsewhere, or they have lacked capital to secure better land, or they were loath to venture into new and unfamiliar areas. Although some good land exists in each broad region of the country, and this reduces the extent of population movements involved in shifting to better land, the adjustment problem is by no means simple.

In the trial-and-error process of adjusting land use to land capability, soil conservation and soil erosion have been only one kind of problem, and often not the most pressing kind. But severely eroded lands usually have become unproductive sooner or later, and ultimately have been abandoned if the process has continued unchecked long enough.

When the U.S. Department of Agriculture undertook the National Inventory of Soil and Water Conservation Needs, data were secured for the first time on a comprehensive, essentially national, scale, which related land capability and land use. The information on land capability was discussed in Chapter 6. Unfortunately, the Conservation Needs Inventory (CNI) included data only in three broad categories— cropland, pastures and ranges, and forests and woodland—and in one miscellaneous class which included mainly idle land, land in roads, and the like. The urban and built-up areas were excluded from the inventory. The latter, while relatively small in acreage, are among the most important lands in the whole country.

There are two sources within the Department of Agriculture for information on acreage used for different purposes—the CNI and figures of the Economic Research Service. The CNI shows 447 million acres of cropland in the forty-eight contiguous states in 1958. The Economic Research Service, building on the 1959 Census of Agriculture, shows 391 million acres used only for crops in 1959, plus another 66 million acres rotated between crops and pasture, or a total of 457 million acres of cropland. This seems to indicate reasonably good agreement between the two sources, at least upon the national level. The CNI shows 485 million acres of pasture and range land, excluding the larger federally owned areas, within the forty-eight contiguous states in 1958; the Economic Research Service, using data from different sources, shows 473 million acres—again, reasonably good agreement. The CNI shows 448 million acres of forest and woodland in the forty-eight contiguous states, outside of the larger federal reservations; the Economic Research Service shows 440 million acres—once again, reasonably close. For our purposes, we neither need to consider the reasons for, or importance of, these differences, nor try to judge which source yields the more accurate figures, to the extent that the apparent differences are real. Our concern is primarily with the relationship between the kind of use and land class, and for this the CNI is adequate.

Of the 447 million acres which CNI reports as cropland, 6 per cent is in land-capability Class I; 43 per cent is in Class II; and 34 per cent is in Class III; or a total of 83 per cent in three classes. (See Table 9.) These lands are suited for continued cropping if the recommended practices are followed—even in the judgment of those men most concerned about erosion hazards. In one sense, it is heartening to find that so much of the cropland meets this test. Another 11 per cent of the cropland is in Class IV. While land of this class is marginal, on the whole, yet some of these lands may be used with good conservation practices, and some—subclasses w (excess water) and s (unfavorable soil)—may have only minor erosion hazards. (The location of the lands used for crops in each of these first four classes has already been shown in Figures 18 through 21, and they were discussed in Chapter 6.)

More than 5 per cent of all cropland, however, is in Class V through Class VIII—lands which definitely are not suited to continued cropping, in the judgment of the soil conservationists. Most of this cropland is in Class VI, the better-grade pastures and grazing land, which is considered submarginal for cropping. One serious aspect is that more than two-thirds of these lower-grade lands used for crops have erosion hazard as their primary limitation. While some of these unsuitable lands used for crops are found in each major region of the country, a relatively larger area is found in the central Great Plains than elsewhere. (See Figure 22 for the location of lands in Class V through Class VII.) Presumably these are lands subject to serious blowing hazard, which should be kept in grass rather than plowed. The same is probably true of the lands shown in the northern Great Plains, but probably most of those in the eastern half of the country are relatively steep lands, subject primarily to a serious water erosion hazard.

The relation between land classification and land use may be looked at from another point of view. Of the 36 million acres of Class I lands in the forty-eight contiguous states, three-fourths are in crops; of the 290 million acres of Class II land, almost exactly two-thirds are in crops; of the 311 million acres of Class III land, not quite half is in

TABLE 9. Area of land in Conservation Needs Inventory according to use in 1958 and land-capability classification

(Millions of acres in 48 contiguous states)

Land classification and subclasses[a]		Land use in 1958				
		Cropland	Pasture and range	Forest and woodland	Other	Total
Class I:	All	27.4	3.9	3.6	1.2	36.2
Class II:	e	99.3	22.8	21.2	6.4	149.8
	w	58.0	9.2	15.6	3.5	86.4
	s	20.9	5.0	6.1	1.1	33.1
	c	14.6	5.8	.2	.3	20.8
Subtotal		192.8	42.8	43.2	11.3	290.1
Class III:	e	95.4	44.9	31.4	6.3	178.0
	w	27.4	10.6	30.4	5.3	73.7
	s	22.3	8.1	15.7	2.1	48.2
	c	7.8	3.0	.1	.1	10.9
Subtotal		152.9	66.5	77.6	13.8	310.8
Class IV:	e	31.9	33.5	26.4	4.0	95.9
	w	5.0	4.9	15.5	2.1	27.5
	s	10.1	13.3	16.1	1.7	41.1
	c	1.9	2.1	.1	—	4.2
Subtotal		48.9	53.9	58.1	7.8	168.7
Class V:	All	1.8	10.5	28.9	1.8	43.0
Class VI:	e	13.6	85.3	45.0	2.9	146.8
	w	.6	3.2	2.7	.7	7.1
	s	3.6	55.8	37.3	1.3	97.9
	c	.1	21.8	3.0	—	25.0
Subtotal		17.9	166.1	87.9	4.9	276.8
Class VII:	e	3.6	71.9	82.9	3.5	162.0
	w	.1	.4	5.3	.9	6.8
	s	1.8	57.3	53.1	1.9	114.2
	c	—	8.9	1.3	1.1	11.2
Subtotal		5.6	138.4	142.7	7.5	294.2
Class VIII:	All	.1	2.5	6.4	17.7	26.7
Total classified area		447.4	484.7	448.4	66.1	1,446.6

[a] The dominant hazards in the subclasses are as follows: *e*, erosion; *w*, excess water; *s*, unfavorable soil conditions in the root zone; and *c*, climate.
SOURCE: From *Basic Statistics of the National Inventory of Soil and Water Conservation Needs*, Statistical Bulletin 317, U.S. Department of Agriculture, August 1962, p. 101.

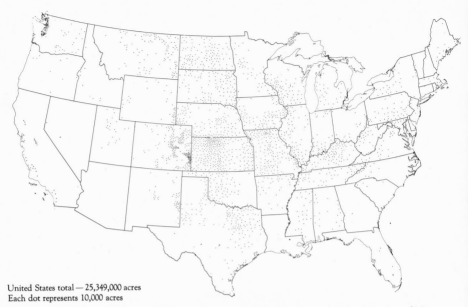

United States total — 25,349,000 acres
Each dot represents 10,000 acres

Figure 22. Land-capability Classes V–VII, 1958 cropland acreage. (Map from Economic Research Service, U.S. Department of Agriculture.)

crops. These relationships may be interpreted in different ways. On the one hand, farmers seem to recognize the inferiorities of the lower-grade lands, and do not use them so fully. On the other hand, not all of the lands of these three classes are in crops. It is true that the capability classification is not directly a productivity classification; yet there must be a high relationship between the two. Productivity enters indirectly because lands are classed for capability with an eye to what farmers have found it practicable to deal with; and also because some of the factors responsible for a high rating in one system are responsible for a high rating in the other. Although Class II and Class III have some deficiencies, in the judgment of the technicians of the Soil Conservation Service (SCS), these lands could be farmed practically and continuously. Most certainly the nation does not need the agricultural output that would result if all Class I through Class III lands were in crops. But this argument is largely meaningless to the individual farmer; for him, prices in the market place are a given factor, they are not influenced by his actions. For him, farm commodity prices are infinitely elastic. If it were profitable to farm these lands at going prices of agricultural commodities, presumably the farmer would put them into crops. Of the Class I through Class III land not in crops, almost half is in forests and woodlands; perhaps it is so costly to clear these lands that it is not worthwhile to do so. But they represent a major cropland reserve. Of the land not in crops in each class, the area in pasture and range is roughly equal to the area in forest and woodland, except in Class V and Class VI—the classes of best pastureland and range lands.

A complete identity of the classification for land use and land capability is impractical and improbable. Some relatively poor land may be included in a field of generally good land and the whole used for crops; or some good land may be included in a field of generally poorer land and the whole used for pasture or woods. Thus, the scale at which the classification is made influences the degree of discrepancy or conformity. A further reason is that some operators have farms too small for full employment of their labor and equipment, and they tend to use poorer lands for crops even when it is not suitable. These and other divergences between farm or field boundaries and soil boundaries account for an unknown, but presumably rather small, part of the total discrepancy between land use and land classification. The greater part is apparently due to the costs of bringing the better grade lands, now used for forests or grazing, into crop production. Clearing, drainage, and other costs may exceed the values that can be created. Minor acreages of good land may also be needed for windbreaks, farmsteads, and other noncrop purposes.

Without attempting to perfect the use of the available data as a measure of productive use of the land, it is possible to identify conformity and nonconformity of capability class and use a little more closely, as shown in Table 10. One example of compatibility of classification and use is croplands of Class I, II, and III; this includes 373 million acres. Another example of conformity, perhaps equally strong, is the use of lower-grade lands for pasture, range, forest, and woodland; 583 million acres are so used. And still another instance of conformity, perhaps not quite so strong, is the use of generally marginal cropland for the same less intensive uses rather than for crops; 112 million acres falls in this group. In total, 1,068 million acres of the 1,447 million inventoried acres in the forty-eight contiguous states were used in conformity with their classification. If the federal agencies are assumed to have brought about a use of the lands under their administration which is compatible with their capability classification, then slightly more than three-fourths of all land is used compatibly.

Various degrees of incompatibility between classification and use can be observed. Perhaps the least serious are the 49 million acres of

TABLE 10. Summary of land use in relation to land capability, 1958

(In 48 contiguous states)

Generalized situation	Land-capability classes	Land use	Area (million acres)
Use and capability compatible:			
Good land, intensively used	I, II, III	Cropland	373
Poor land, extensively used	V, VI, VII, VIII	Pasture, range, forest, woodland	583
Land of dubious quality, extensively used	IV	Pasture, range, forest, woodland	112
Subtotal			1,068
Use and capability incompatible:			
Land of dubious quality, intensively used	IV	Cropland	49
Good land, extensively used	I, II, III	Pasture, range, forest, woodland	238
Poor land, intensively used	V, VI, VII, VIII	Cropland	25
Subtotal			312
Total			1,380

SOURCE: Table 9.

generally marginal croplands (Class IV) used for crops; perhaps next most serious are the 238 million acres of good cropland (Classes I through III) used for pasture, range, forestry, or woodland uses; and the most severe incompatibility is the 25 million acres of poorer classes of land used for crops. One might take some satisfaction in the fact that the acreage of good land used less intensively than it could be is greater than the acreage of poor land used too intensively. But 25 million acres of poor land used too intensively is hardly a cause for satisfaction.

Not all of the compatibly used lands have adopted good conservation practices, but they all could do so. To use Class III croplands properly might require changes in crop patterns and sequences, or in tillage practices, or the construction of works such as terraces, or other special measures. It might not pay the farmer to adopt these special practices, or he might not regard them as necessary, or he might not be convinced of their efficacy or profitableness. But it would be possible to achieve conservation on these lands without major shifts in land use, from crops to grass or trees.

In this connection, it would be interesting to know how much of the poorer land used too intensively was found in whole farms and how much consisted of small fields within farms which contained better

TABLE 11. Land classification of major land uses along a specific traverse in Kit Carson County, Colorado, 1939–1940 and 1962

Land use	Major land-capability classes as percentage of total			
	III	IV	VI	VII
1939–40:				
Cropland	12	68	16	3
Idle land	18	60	17	5
Pasture	10	52	25	13
1962:				
Wheat land	16	68	13	3
Other cropland	22	50	23	5
Conservation Reserve	8	49	34	9
Abandoned land	0	67	26	8
Pasture	8	51	27	14

SOURCE: Adapted from Leslie Hewes, ''A Traverse across Kit Carson County, Colorado, with Notes on Land Use on the Margin of the Old Dust Bowl, 1939–1940 and 1962,'' *Economic Geography*, Vol. 39, No. 4 (October 1963).

land. But such information is not available from the Conservation Needs Inventory.

A recent study by Leslie Hewes of one locality in eastern Colorado employed the SCS land-capability classification and related shifts in land use over the past twenty years to this classification.[1] At the start of the period studied, in 1939–40, the area showed relatively little conformance of land use to land capability. (See Table 11.) The best land in the locality was Class III on the SCS land-capability classification scale. Relatively more of idle land than of cropland was in Class III, and pastureland accounted for nearly as large a percentage as cropland in this class. A substantial part of the cropland was Class VI or worse. Much of this land had been blown seriously in the immediately preceding years and, of the blown land, much was idle.

During the intervening twenty years, much of the previously abandoned land was again plowed, largely under the stimulus of relatively profitable wheat prices. The war years were generally years of favorable rainfall as well as of favorable prices. The early fifties saw new droughts, however, and this locality experienced some of the soil blowing shown in Figure 7 and Figure 8. Nevertheless, the land was in better physical shape on the whole in the later than in the earlier years. Hewes says of the later period, "It seems sufficient here to call atten-

---

[1] Hewes, "A Traverse across Kit Carson County, Colorado, with Notes on Land Use on the Margin of the Old Dust Bowl, 1939–1940 and 1962," *Economic Geography*, Vol. 39, No. 4 (October 1963).

tion to the remarkable 'come back' of the land and farmer and to
suggest that there is little to indicate any widespread retreat of agricultural occupance. However, the soil-banking of land . . . suggests a
measure of failure." More pertinent to our present purpose, land use
in 1962 was somewhat better adjusted to land capability than it had
been previously. Despite considerable divergence which remained,
more of the better land was in crops, fewer crops were grown on
unsuitable land, and generally there was a better conformance of land
use to land character. Hewes comments on the situation: "The fact
that the contrast in land class by land use is not very sharp—that some
land of all four classes [III, IV, VI, and VII] is used for crops, both for
wheat and other crops, [and some] is in pasture, and has been put in
the Soil Bank—suggests that individual decision is important." In spite
of these adjustments toward better land use, this is still a very marginal
agricultural region, which over the decades has depended heavily upon
federal subsidy in the form of supported wheat prices and in other
ways. It has been undergoing profound changes in numbers and size
of farms and of farm population. However, it is heartening to learn
that, in this one area at least, shifts in land use have been in the
direction of better land use.

In considering the degree of conformity between land use and its
classification, it is helpful to recall the processes of land-use adjustment. A farmer can rather readily adjust land use to land capability
within the limits of his farm; he may be limited by his knowledge of
the land capabilities, or by the availability of capital to carry out
changes, or by the costs of clearing land and making other changes
which make adjustments of dubious profitability, or by other factors.
Yet such obstacles may not be too serious, and there are almost always
major advantages in using land according to its capabilities. From the
same inputs, the farmer will gain a larger gross return, and hence a
much larger net return, if land is used to its best capabilities.

These relationships do not hold as well for farms in the same community or county. Through the operations of the land market, some
farmers can buy land and adjust its use to the best capabilities. But
the area of land which changes hands annually is small, and many
sales and purchases largely preclude any interfarm changes in use.
There are few effective mechanisms at the local or county level for
the achievement of land-use changes. There are planning and zoning
powers for rural areas in a few states which have more or less active
programs; but in most parts of the country either the legal powers are
lacking or there are no local or county programs in operation. The U.S.

Department of Agriculture and the land-grant colleges embarked upon a land-use planning program in the later part of the thirties which had as one of its purposes this type of interfarm adjustment in land use. But that program was not in operation long enough to become very effective; perhaps it could not have been in any case.

There are still fewer programs for land-use adjustment when one moves up to the state and regional levels. No state, as far as we know, has any program for planning, zoning, or otherwise seeking to promote or facilitate land-use changes between counties or other major parts of the state. The federal agricultural programs of the last thirty years have scarcely touched the problem of interregional shifts in land use; and, through the use of historical bases for crop control programs and in other ways, actually may have impeded, rather than facilitated, such adjustments. Regional shifts in crop production have occurred. As some regions have lost out in the competitive race, farms have been abandoned or combined with other farms. In considerable part, such adjustments have occurred as farm operators grew old, retired, or died. It has been, on the whole, a slow and haphazard method of land-use adjustment.

It might be argued that the difficulties of planning and implementing programs to facilitate land-use changes among farms, counties, and regions are too great, or the dangers from centralized planning are too serious, for any public program to be undertaken toward this end. But substantial arguments could be marshaled on the opposite side. Whereas we do not wish to pursue this issue at this time, we think it is necessary to recognize the obstacles to adjustment and the lack of mechanisms to that end.

Chapter 8

MEASURES OF PROGRESS ON
SOIL CONSERVATION
SINCE 1930

THE proof of the pudding is in the eating; the time has come to eat. What have been the effects of the various soil conservation programs, public and private, since about 1930? The year 1930 as a reference point is used because, in spite of the difficulty of identifying any wholly satisfactory starting date for a process of historical change, that was just before the relatively large-scale soil conservation programs of the early New Deal days.

There is fairly general agreement that substantial progress has been made in soil conservation in the past generation or so. But just how much progress has actually been made? Perhaps a prior question is: How does one best measure "progress" in this field? This, in turn, raises a still earlier question: What is considered a satisfactory state of soil conservation—just how much erosion, if any, is the country willing to accept? How does one measure progress from any given position toward such a desired or satisfactory level? These questions are far from simple; equally competent and equally honest men may come up with different answers, depending in part upon their particular viewpoints but also largely upon which facts they put foremost.

When it comes to measuring progress, we have several sources of recent information: (1) the 1957 National Inventory of Soil and Water Conservation Needs (CNI) lists areas not needing conservation treatment and areas already adequately treated; (2) substantial amounts of public funds have been expended under the Agricultural Conservation Program (ACP) of the Agricultural Stabilization and Conservation

Service (ASCS), and a record of the practices carried out gives some indication of the amount of soil conservation achieved thereby; (3) the 1959 Census of Agriculture contains information on certain soil conservation measures; (4) a number of special studies have been made in localized areas which throw some light on soil conservation progress; and (5) workers in agriculture have some knowledge, or at least a consensus, on certain points which may have value. Each of these sources of information has its special characteristics. In addition, while direct comparison is impossible, some estimate of progress can be made by comparison with the Reconnaissance Erosion Survey of 1934.

Despite the diversity of sources, there are many inadequacies in data relating to soil conservation achievements; one reason may be connected with the general history of soil conservation efforts. Prior to 1930, there was both a widespread indifference to the problem of soil erosion and a great ignorance about it. There was very little quantitative data, and none on a national scale. With the coming of the New Deal, a massive national program of soil erosion control was launched immediately; there was little time or manpower available for careful measurement of either situation or progress. Moreover, the crusading character of that program was significant: soil erosion is taking place; it is bad; therefore, move with all possible energy and speed to cure it, wasting time in neither contemplation nor unnecessary detail. Whereas some measurements were made from the beginning, these aspects of the problem have never had as much attention as the action aspects.

Assessing changes in the national soil resource situation is a difficult job—partly because little information is available for the period prior to the thirties to provide a base from which to work. The job is less arduous today than a generation ago simply because of the progress made in the study of soils and water and of their interaction under widely varying conditions. A systematic and thorough appraisal of soils can be undertaken today even though the mapping of the nation's soils is far from finished. An increasing amount of information is becoming available on the relative erosion hazard on different soils but there is little on the current rate of soil loss taking place on them.

No one has undertaken an analysis for the country as a whole which would show the relative economic impact of additional soil erosion or the long-run economic gains from additional sums spent for more erosion control than that now in effect. Much more information is available as to the physical situation on the land, but even here the statistics leave much to be desired. The federal agencies report areas covered by their programs. In no case is this the entire farm acreage since some farms are not involved in any program, but the divergence probably depends upon the particular soil conservation practice. Where the cost is relatively large, requires fairly high technical skill, and is applied only once—terracing is an excellent example—the probabilities are that most of the land is included in one of the federal programs. Where the practice is simple and must be applied each year—contour plowing and cultivation, for example—it is probable that many farmers undertake it on their own, or with technical help from nonfederal public agencies or private organizations. The data from the federal agencies for the area covered each year are reported differently: the ACP, for example, often includes areas which have been reported under the same practice in earlier years; while the Soil Conservation Service (SCS), at least in recent years, has reported the area of accomplishment.

Other difficulties also exist. Although conservation practices may be applied, the way in which they are done may not be adequate technically and the resulting soil conservation may be less than could reasonably have been achieved. With increasing competence at all levels, public and private, this is almost certainly less common today than formerly, but there is still imperfection. More serious, soil conservation is not something done once and then forgotten; much of it has to be repeated. Annual maintenance may be needed, or an annual decision to repeat the conservation practice. The available statistics also include many practices whose effectiveness for soil conservation may be doubtful. Application of fertilizer to establish grass to prevent erosion may be a soil conservation practice, for instance, but application of fertilizer to increase crop yields generally would not be. The same distinctions can be made for the application of limestone and of several other practices.

The Situation in 1930

If it is difficult to know with any degree of detail what the situation with respect to the condition of soil resources is today, it is obviously even more difficult to define the situation in 1930. Certain available pieces of information, however, indicate some of its dimensions in a crude way. The agricultural experiment station at Spur, Texas, at that time had an experimental plot on which data on erosion losses had been collected since 1926. After the congressional authorization and appropriation for similar work on a national scale, more data began to be collected. These efforts did not get under way before 1930 and it was a number of years before the results became available. Table 12 presents some of the research findings as they appeared in 1939; but these fairly well reflect the actual situation as of 1930.

The rates of soil loss observed by these stations under experimental conditions indicated what could be expected given either of two relatively extreme cases: continuous cropping to row crops, or continuous sod or meadow crops. Valuable as this information was, with-

TABLE 12. Soil losses under cultivation and protective cover for thirteen soils representative of erodible farm land for an area of 250 million acres, 1939

Soil type, location and year of measurement	Annual rainfall (inches)	Slope (per cent)	Clean tilled		Dense cover	
			Crop	Annual soil loss per acre (tons)	Crop	Annual soil loss per acre (tons)
Cecil sandy clay loam Statesville, N.C. 1932–36	48.42	10.0	Cotton	25.08	Mixed grasses	0.014
Nacogdoches fine sandy loam Tyler, Tex., 1932–36	42.34	10.0	Cotton	8.18	Bermuda grass	.006
Kirvin fine sandy loam Tyler, Tex., 1932–36	41.77	8.75	Cotton	30.07	Bermuda grass	.05
Shelby silt loam Bethany, Mo., 1931–35	34.79	8.0	Corn	68.78	Alfalfa	.25
Muskingum silt loam Zanesville, Ohio, 1934–36	36.46	12.0	Corn	72.23	Bluegrass	.04
Clinton silt loam La Crosse, Wis., 1933–35	34.12	16.0	Corn	88.66	Bluegrass	.03
Dubuque silt loam La Crosse, Wis., 1934–35	36.53	30.0	Corn	81.44	Bluegrass	.22
Vernon fine sandy loam Guthrie, Okla., 1930–35	33.12	7.7	Cotton	24.29	Bermuda grass	.032
Houston Black clay Temple, Tex., 1931–36	32.76	4.0	Corn	23.83	Bermuda grass	.03
1933–36	34.90	2.0	Corn	10.62	Bermuda grass	.10
Marshall silt loam Clarinda, Iowa, 1933–35	26.82	9.0	Corn	18.82	Bluegrass	.06
Palouse silt loam Pullman, Wash., 1932–35	21.74	30.0	Wheat, Fallow	8.52	Mixed grasses	1.45
			Bare	27.82	Spring wheat	1.65
Colby silty clay loam Hays, Kan., 1930–35	20.36	5.0	Kafir	11.74	Native grass	.029
Miles clay loam Spur, Tex., 1926–37	20.73	2.0	Cotton	7.03	Buffalo grass	1.29

SOURCE: From U.S. Department of Agriculture, *Agricultural Statistics, 1939* (Washington, Government Printing Office), p. 530.

out an accompanying report of the prevalent cropping practices within each of the areas for which these soils were typical, there is no way to know what the soil loss rate was under actual farming conditions at that time.

The production of row crops entails a greater erosion hazard than that involved in the production of small grains or hay. Approximately 43 per cent of the cropland that was harvested in the period 1957–61 was devoted to such major row crops as corn, sorghum, cotton, and tobacco. The same proportion of cropland was used in this way during the five-year period centering on 1930. Erosion losses could have been greater in the earlier period, however, if for no other reason than the fact that the twenties and thirties were decades in which the total acreage of cropland in the nation was at peak levels. For the five-year period centered on 1930, nearly 357 million acres of crops were harvested. In contrast, the average harvested acreage for the years 1957–61 was only 312 million acres.

It can also be assumed that soil losses were greater in the earlier period because of the extremely limited effort then being made to deal with the situation. Contouring and terracing were relatively unknown in the United States. Tillage was relatively excessive by today's standards which stress the minimum disturbance of the soil necessary to obtain a satisfactory seedbed and the control of weeds. Crop rotations had no place in the operation of many farms—particularly those which were primarily devoted to the production of a single crop, as was often the case in cotton areas. Farms with little or no livestock had little, if any, cropland pasture or meadow crops in their rotations and there was often a tendency on those farms with livestock to keep the same fields for hay and pasture year after year rather than rotating them with row crops.

A calculation in the original table, from which data in Table 12 were taken, indicated the number of years that would be required at various rates of soil loss in order to remove the soil to a depth of 7 inches from the surface. At LaCrosse, Wisconsin, where a soil loss rate of over 80 tons per acre per year was observed on the experimental plots that were planted continuously to corn, the 7 inches of soil would have been removed in only 11 years. If these same soils had remained forever in bluegrass, it would have taken a period of over 4,000 years to remove the same amount of soil from the fields with the steepest slopes. On the equally steep slopes of the Palouse silt loam soils in the state of Washington, under plantings of spring wheat or mixed grasses, or on the Miles clay loam at Spur, Texas, where buffalo grass was the cover, the 7 inches would have been lost in a much shorter time, from 650 to 830 years, even though not planted to row crops. In some situations, however, the erosion losses from soils with a dense protective cover were so slight that as many as 214,200 years would be required to dislodge 7 inches of soil. Offsetting soil losses are the additions to topsoil which occur as natural processes slowly convert subsoil materials into true soils. These processes proceed at a slow rate but generally would offset, or more, soil losses reported under close-growing cover.

Although measurements of soil loss by soil type and slope class under various crop rotations were sorely needed, they were not generally available even in the late thirties. The available data consisted at best only of a scattered sample, yet the national large-scale soil conservation program necessarily had to go forward during these years on the basis of such information as there was.

The Reconnaissance Erosion Survey of 1934 and Its Follow-Up

One of the early tasks undertaken by the Soil Erosion Service under Hugh Hammond Bennett's direction, we noted in Chapter 4, was a quick Reconnaissance Erosion Survey of the entire country. Until that time, no one had really known how extensive the erosion problem was. In undertaking this survey, Bennett was encouraged by Secretary of the Interior Harold L. Ickes and the National Resources Board. The survey was made during 1934, by 115 trained soil erosion specialists, in only two months' time. Done hurriedly, it had some serious deficiencies; yet it was a landmark, not only because it was the first national survey, but also because it attempted to classify the kinds of erosion and to measure the degrees of severity of erosion. In these respects it was much superior to more recent surveys which, in spite of taking much longer time with more manpower and having better basic knowledge from which to proceed, have produced basically less usable information. The 1934 survey was criticized at the time, and since, for exaggeration, for failing adequately to recognize that some erosion is geologic and unaffected by man, and for not recognizing that some areas are growing by deposition rather than losing by erosion. Nevertheless, the results have stood up remarkably well.

The first map from this survey—and the first map showing soil erosion on a national basis and with comparable definitions used for the entire country, as far as we can learn—appeared in the December 1934 report to the President by the National Resources Board (which had succeeded the National Resources Committee and which was shortly to be succeeded by the National Resources Planning Board). We have presented a reproduction of part of that large and detailed map in Figure 6. A very slightly revised version of that first map appeared in *The Yearbook of Agriculture, 1938*,[1] and is reproduced here as Figure 23. Some of the extreme language of the first version was dropped—for example, "essentially destroyed" in describing the most severe wind and water erosion—and there were modest changes in the boundaries of many areas, especially in the West. Still another version appeared in *The Yearbook of Agriculture, 1957*,[2] which we reproduce as Figure 24. This map has only three broad classes—slight or none, moderate, and severe—in contrast to the more detailed but still generalized classification in the two earlier versions. Some boundary lines have been changed, and some areas have been rated as more severe in the recent version while others are less so than earlier. Perhaps more striking than the differences among the three maps, however, are the broad similarities. While time has improved knowledge, as well as perhaps changing the basic situation, the earlier results, hastily assembled though they were, have stood up reasonably well.

The same information was also presented in statistical form. The first publication of these data was in the December 1934 report to the President by the National Resources Board, and they are reproduced here in Table 13. Within a few months, somewhat refined and revised data were published as part of the report of the Land Planning Committee of the Board, and these, too, are included in Table 13. The part of each report dealing with soil erosion was prepared by Bennett and his aides. Differences in the two reports thus reflect their revisions of their own data; while these revisions were rather large—probably resulting partly from the extreme haste in which the Reconnaissance Survey had been made and the lack of adequate basic data at the time—the picture is similar for the two. The first report used an incorrect figure for total national land area, and the area shown in several classes differs by many millions of acres between one report and the other.

[1] U.S. Department of Agriculture, *Soils and Men—The Yearbook of Agriculture, 1938* (Washington: Government Printing Office).
[2] U.S. Department of Agriculture, *Soil—The Yearbook of Agriculture, 1957* (Washington: Government Printing Office).

Results of the Reconnaissance Soil Erosion Survey of 1934 were presented in terms of about thirty descriptive categories for those areas where accelerated erosion, in contrast to geologic erosion, was found. The system differentiated among areas on the basis of the proportion of the top soil that had been lost, and also among those which had suffered from sheet erosion, gullying, wind erosion, or some combination of these. An area in which less than 25 per cent of the top soil had been lost was regarded as having little or no erosion; the range between a loss of 25 per cent and 75 per cent was considered a moderate degree of erosion; and losses greater than 75 per cent were classified as severe. Compound descriptions, such as "moderate sheet erosion and occasional gullies" and "slight sheet erosion and frequent gullies" and "moderate sheet erosion, severe wind erosion and occasional gullies" were used to distinguish types and degrees of damage. The survey was, as intended, nothing more than a reconnaissance effort. It permitted broad estimates to be made of the situation and prompted the observation that "erosion over the United States . . . is much more widespread and more severe than had been previously estimated." It carried the implication that the situation was growing worse rather than becoming better, but it provided little information on the historic situation, or the trends in erosion. Current cropping practices and the level of soil management were noted as indicators of what was then happening, but these observations were not quantified systematically. Thus, while the national summary gives a picture of damage which had taken place over the years until the time of the survey, little evidence was presented to support the contention that the situation was tending to become worse—although that may well have been the truth.

The data from Table 13 are summarized further in Table 14. Roughly one-half of the national land area can be said to have had only slight or no erosion; while much of this may have suffered needless damage, or may not have been used as efficiently as possible, on the whole this was not a serious problem. Approximately half had moderate to severe erosion; some of the latter was very bad, with easily obtainable horror pictures showing man-made gullies that were miniature Grand Canyons or man-made sand dunes of impressive dimensions. However, of the worst half of the land, nearly three-fourths was in the moderate category. Only about one-seventh of the national area was in the worst categories. The optimist can see in these figures a situation not too desperately bad, while a pessimist

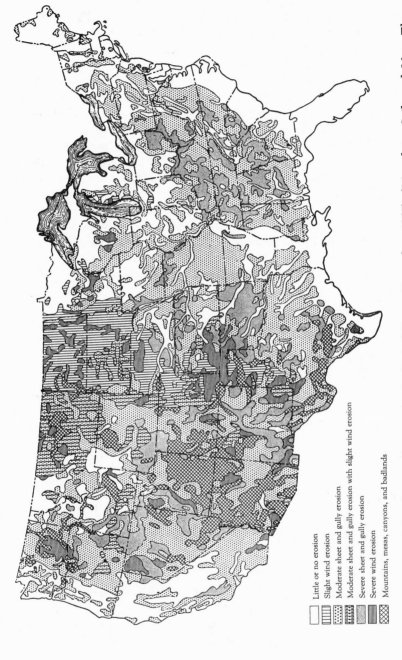

Figure 23. General distribution of erosion in the United States, December 1937. (Map from *Soils and Men—The Yearbook of Agriculture, 1938*, p. 93.)

Little or no erosion
Slight wind erosion
Moderate sheet and gully erosion
Moderate sheet and gully erosion with slight wind erosion
Severe sheet and gully erosion
Severe wind erosion
Mountains, mesas, canyons, and badlands

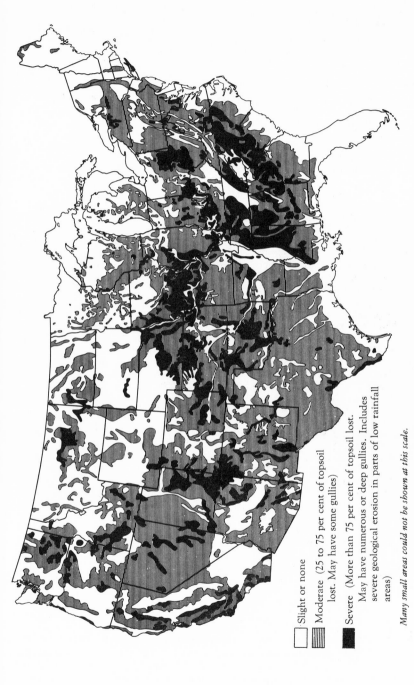

Slight or none

Moderate (25 to 75 per cent of topsoil lost. May have some gullies)

Severe (More than 75 per cent of topsoil lost. May have numerous or deep gullies. Includes severe geological erosion in parts of low rainfall areas)

Many small areas could not be shown at this scale.

Figure 24. Generalized soil erosion, October 1948. (Map from *Soil—The Yearbook of Agriculture, 1957,* p. 307.)

TABLE 13. Reconnaissance Erosion Survey of 1934, as described in two reports of the National Resources Board

(Millions of acres)

	Data from Report of December 1, 1934		
Code Number	Description of class[a]	Total area	Per cent of national area
1	Practically no erosion (best cropland, much mountain forest)	531	27.4
17	Little sheet erosion, some gullying	137	7.1
18	Little sheet erosion, severe gullying	19	1.0
2	Moderate sheet erosion (early stages of decline)	100	5.2
27	Moderate sheet erosion, some gullying	252	13.0
28	Moderate sheet erosion, severe gullying	242	12.5
3	Severe sheet erosion (productivity greatly reduced)	8	.4
37	Severe sheet erosion, some gullying	59	3.0
38	Serious sheet and gully erosion	128	6.6
4	Moderate wind erosion (best western farm and grazing land)	131	6.8
47	Moderate wind erosion, some gullying	18	.9
48	Moderate wind erosion, severe gullying	3	.2
24	Moderate sheet and wind erosion (good grazing, fair cropland)	13	.7
47	Moderate sheet and wind erosion, some gullying	30	1.6
248	Moderate sheet and wind erosion, severe gullying	29	1.5
348	Serious sheet, moderate wind erosion, severe gullying	8	.4
5	Severe wind erosion (largely submarginal for cultivation)	55	2.8
57	Serious wind erosion, moderate gullying	4	.2
58	Serious wind erosion, severe gullying (mainly valueless)	[c]	[d]
25	Moderate sheet, serious wind erosion (no gullies)	8	.4
257	Moderate sheet, serious wind erosion (some gullying)	6	.3
35	Serious wind and water erosion, no gullies (most of little value)	2	.1
357	Serious wind and water erosion, some gullying (most of little value)	6	.3
358	(This class not shown)		
6	Severe wind erosion (most essentially destroyed)	9	.5
67	(This class not shown)		
68	(This class not shown)		
9	Essentially destroyed, most by gullies (formerly cultivated)	13	.7
9A	Mainly badlands and stony lands in West	112	5.8
24W	Chiefly scablands, locally subject to blowing	6	.3
R	Largely bare above timberline (mainly valueless)	5	.3
347	(This class not shown)		
258	(This class not shown)		
	Total	1,935	100.0

[a] Because of limitation of space, this description is somewhat abbreviated from original. It has been done mainly by dropping off descriptive evaluations of usefulness of each soil.
[b] This description has been included with almost no change from the original.
[c] Less than 500,000 acres.
[d] Less than one-tenth of 1 per cent.

TABLE 13.—(continued)

(Millions of acres)

	Data from Report of 1935		
Code Num- ber	Description of class[b]	Total area	Per cent of national area
1	Little or no erosion	576	30.3
17	Slight sheet erosion, occasional gullies	94	5.0
18	Slight sheet erosion, frequent gullies	17	.9
2	Moderate sheet erosion	98	5.1
27	Moderate sheet erosion, occasional gullies	296	15.6
28	Moderate sheet erosion, frequent gullies	174	9.2
3	Severe sheet erosion	7	.4
37	Severe sheet erosion, occasional gullies	61	3.2
38	Severe sheet erosion, frequent gullies	109	5.7
4	Slight wind erosion	124	6.5
47	Slight wind erosion, occasional gullies	16	.9
48	Slight wind erosion, frequent gullies	1	.1
24	Moderate sheet, slight wind erosion	19	1.0
47	Moderate sheet, slight wind erosion, occasional gullies	43	2.3
248	Moderate sheet, slight wind erosion, frequent gullies	17	.9
348	Severe sheet, slight wind erosion, frequent gullies	9	.5
5	Severe wind erosion	55	2.9
57	Severe wind erosion, occasional gullies	6	.3
58	Severe wind erosion, frequent gullies	c	d
25	Moderate sheet, severe wind erosion	6	.3
257	Moderate sheet, severe wind erosion, occasional gullies	3	.2
35	Severe sheet and wind erosion	c	d
357	Severe sheet and wind erosion, occasional gullies	c	d
358	Severe sheet and wind erosion, frequent gullies	2	.2
6	Very severe wind erosion	9	.5
67	Very severe wind erosion, occasional gullies	c	d
68	Very severe wind erosion, frequent gullies	c	d
9	Destroyed by gullying	4	.2
9A	Mesas, canyons, badlands, rough mountain land	130	6.9
24W	Scablands, shallow soils, moderate sheet and wind erosion	10	.5
R	Bare mountain tops	5	.3
347	Severe sheet and slight wind erosion, occasional gullies	4	.2
258	Moderate sheet and severe wind erosion, frequent gullies	7	.4
	Total	1,903	100.0

SOURCES: Description and data for 1934 adapted from National Resources Board *Report on National Planning and Public Works in Relation to Natural Resources and Including Land Use and Water Resources, with Findings and Recommendations* (Washington: Government Printing Office, Dec. 1, 1934), Table 17, p. 168. That for 1935 is from *Supplementary Report of the Land Planning Committee to the National Resources Board*, Part V, "Soil Erosion: A Critical Problem in American Agriculture" (Washington: Government Printing Office, 1935), Table VIII, p. 23.

TABLE 14. Generalized soil erosion situation, 1934

General degree of soil erosion[a]	Code numbers of classes included	Total area	
		Million acres	Per cent of national area
Little or none	1	576	30.3
Slight	17, 18, 4, 47, 48	252	13.3
Moderate	2, 27, 28, 24, 247, 248, 25, 257, 258	663	34.8
Severe or serious	3, 37, 38, 347, 348, 358, 5, 57, 6	262	13.8
Destroyed by gullying	9	4	.2
Miscellaneous	9A, 24W, R	145	7.6
Total		1,903	100.0

[a] Grouped according to descriptive word for first type of erosion listed in each class; some have severe secondary types of erosion. For instance, class 48 (slight wind erosion, frequent gullies) is included as "slight."
SOURCE: Table 13.

could easily project a few generations forward to national impoverishment because of a scanty soil base.

The 1934 Reconnaissance Survey attempted no measure of the productive potential still possessed by the different eroding soils. This potential, while partially indicated by the proportion of the top soil reported to have been lost, would also have required knowledge of the response of the top soil remaining to appropriate soil management practices, as well as an understanding of the qualities and characteristics of the entire soil profile. This information was not then available.

The findings indicated that serious sheet erosion in the southeastern, southern, and the southwestern-central parts of the United States had caused "formerly productive" agricultural land to be abandoned. No estimate of the former level of productivity of these soils was made. Some of them undoubtedly had never had a high level of natural fertility. The depletion of this store of plant nutrients may have come not many years after the soils were first brought into cultivation. Unless measures were taken to maintain the fertility of such soils, the erosion hazard would have increased as the crops growing there became less vigorous and afforded less protection against sheet erosion. Off-farm employment opportunities, or out-migration, solved the problem after a time in some of the areas where soils had become depleted. The lands dropped out of cultivation and in a matter of time were protected against erosion by the establishment of new natural cover.

Land abandonment was much less frequent in the hill regions of the southern states and the Ozarks. The rate of natural population increase was high and out-migration was too low to offset it. Lack of economic opportunities and other factors caused the residents of these areas to continue to attempt to farm soils which had long been at a disadvantageous competitive position with the better endowed, level lands elsewhere where mechanized production was feasible. Continued cultivation of the poor hill land was a losing proposition for those farming it, as the soil itself began to wash away. Abandonment would have been the best thing that could have happened to such areas, but it was slow in coming about where those who were using such soils had limited alternatives, or none at all.[3]

An area of 663 million acres was found to have lost between one-fourth to three-fourths of the topsoil. Although the portion of this which was agricultural land was still deemed to be good, it was reported to be "rapidly losing additional topsoil." Gullying also was widespread. In nearly nine cases out of ten, it was associated with sheet erosion. Of the 865 million acres on which Table 13 shows that there was gullying in some degree, the gullies were judged to be most severe on 337 million acres. Slightly more than 4 million acres were said to have been destroyed in this way. The gullies found in the arid regions, although largely the result of geological processes, were reported often to have been accelerated by the manner in which the land had been used. Wind erosion was found to have affected a total area of 322 million acres, most of which lay within the Great Plains, from the Dakotas down through southwestern Texas. Of this acreage, 80 million acres were judged to have been seriously damaged. Another 9 million acres were, from a practical standpoint, considered to have been destroyed.

Past erosion in some areas had removed substantial amounts of soil, according to the survey's findings. Whether these losses had occurred much earlier and were tending to diminish when the survey was made, or whether they were continuing unabated, the survey does not always indicate. Certainly there must have been many areas in which the problem, particularly if it were primarily one of sheet erosion, continued with little or no awareness of it by owners or farm operators. On the other hand, changes were under way in 1934 that must have been slowing to some extent the rate of loss on some soils.

The amount of fertilizer use in 1928, 1929, and 1930 was about 1.5 times greater than in 1910; although this rate of use was at best only a

[3] C. P. Barnes, "The Problem: Land Unfit for Farming in the Humid Areas," in *Soils and Men, op. cit.*, pp. 60–67.

sixth of the current rate of use and concentrated in a small part of the nation, it must have made some contribution to erosion control. The use of green manures and animal manures must have increased somewhat over the years, and while none of these measures alone would have been sufficient to prevent erosion losses, anything which tended to promote more vigorous plant growth would have made some contribution. Probably, however, it was not the farms on which the erosion hazards were the greatest or the farms with the poorest soils on which the fertilizers were coming into use. More farmers were beginning to appreciate the importance of returning crop residues to the soil, but up to 1940 fires in the spring were seen throughout the Corn Belt when farmers were burning the past year's dried corn stalks to get rid of them before seeding the new oat crop.

The Situation in 1958 as Revealed by the Conservation Needs Inventory

In the years following the 1934 Reconnaissance Soil Erosion Survey, the Soil Conservation Service, other agencies, and farmers learned a great deal about soil erosion, its control, and land management in general. Experimental results began to accumulate, the large-scale practical operating programs taught many lessons, and there were many more trained and experienced men than there had been in 1934. Knowledge accumulated in many places.

But no comprehensive survey of soil erosion was undertaken after the 1934 effort. In large part, this was because the emphasis in the soil conservation program had shifted. No longer did SCS consider the control of soil erosion as its sole or chief job. Its original purpose was to preserve the existing basic productivity of the soil, and for this it had to have the best possible estimate of the hazards to that productivity. By the late forties, SCS had shifted primarily to an emphasis upon land management for greater output and for higher income, of which erosion control was only one part. There seemed to be less need for a close estimate of the nature and severity of the soil erosion problem.

It was under the leadership of SCS that the Department of Agriculture undertook the Conservation Needs Inventory in 1957. As we pointed out in Chapter 6, one basic strength of the CNI was the careful and scientific sampling of areas from which data on land capability and other matters was obtained. Of the four basic jobs assigned to this Inventory by the Secretary of Agriculture, the first (to measure land capability) should have obtained information on current soil erosion, on the tendency of the land to erode, and on past erosion conditions, since these factors enter into the classification of land according to land capability. If such data were obtained by the men on the location, however, they were completely submerged in the subsequent analysis and publication of the information. Table 8 in Chapter 6 shows the area of Class II soils on which erosion is the dominant problem; from the definition of this class, it might reasonably be assumed that erosion was slight on these soils. But some Class III and IV soils, in which other limitations are dominant, may well have slight erosion of the same degree as the Class II soils. Some inferences as to degrees of erosion can be drawn from data such as are in Table 8, but they are inferences only. They do not give any hint as to whether erosion is caused primarily by wind or water, and they provide no direct information as to the trend in erosion in the past. The Inventory is not an objective measurement and presentation of the facts about the soil erosion which existed then and its trends, but is primarily the conclusions of the U.S. Department of Agriculture (and mainly of SCS) or its interpretations of facts which have not been published in a form permitting others to test these conclusions or to make other analyses.

As part of CNI's assignment to determine what conservation treatments were needed, it was necessary to estimate the lands not needing conservation treatments as of January 1, 1958. The men who carried out the CNI at every level were deeply conscious of the soil conservation problem, and they surely did not underestimate the magnitude and seriousness of the soil conservation job still to be done; in fact, they may have overestimated it, as others would have judged the job. It is therefore possible that they somewhat underestimated the extent of the accomplishments up to 1958; at least, it seems unlikely that they overstated them. Also, in carrying out instructions for estimates based upon the expected crop acreage of 1975, CNI had to distinguish areas which did not need treatment or were not feasible to treat. The statistical tabulations from the CNI provide somewhat more detail than one would have expected from these instructions, in that for the cropland category the areas not needing treatment are separated from those already adequately treated.

Some of the cropland has no significant soil conservation problem. The instructions stated that Class I land was to be placed in this category. In fact, the data follow this instruction very closely, but not exactly, on a state-by-state basis, as shown in Figure 25. Divergences in area from this relationship were not great. In addition, some 136 million acres of land in the United States, expected to be used for cropland in 1975, were considered to have been adequately treated before 1958. (See Table 15.) These lands may have been of any class except I, which was excluded by definition as not requiring treatment. If soil conservation treatments had been confined to lands capable of permanent cropping (Class II through Class IV) and also needing conservation measures, then this area of adequately treated lands can

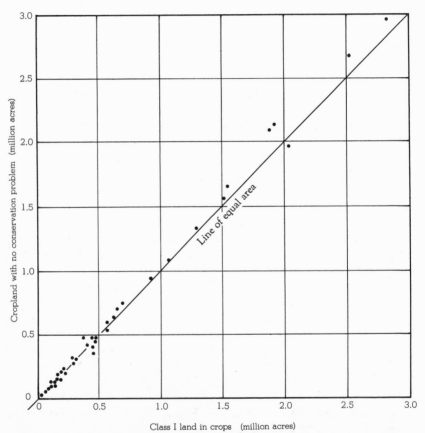

Figure 25. Relation between acreage of Class I lands in crops and cropland with no conservation problem, by states, 1958. (Note: It is impossible to show each state by a separate dot, especially where acreages are small; hence, some states had to be omitted.)

be compared with the acreage in Class II and Class III or in Classes II through IV. On a national basis, 39 per cent of the Class II and Class III land expected to be used for crops in 1975 had been adequately treated by 1958. The figure is 34 per cent of the area if Class IV land is also included. The various states are rather close to this national pattern, with over two-thirds of the forty-eight contiguous states falling in the range of 30 per cent to 39 per cent of the Classes II through IV land so treated.

No information was provided in the CNI as to the kinds of treatment that had been carried out on these lands, the costs of such treatment, the degree to which treatment had been aided by public programs or had been entirely by private action, or as to the severity of the conservation problem before treatment. Were the most severely eroding lands treated, or were the lands easiest to treat actually cared for, or what was the situation? It could well be argued that farmers and public agency people alike would wish to concentrate their efforts on the most seriously eroding lands. If so, treatment of the worst 34 per cent of the cropland area would be a great achievement. On the other hand, treatment of only slightly eroding areas is so much easier, that perhaps much of the treatment was done there; if so, the achievement of 34 per cent is much less impressive. Moreover, treatments need not have been confined to lands with an erosion problem; in keeping with the general approach of the CNI, they might have included lands drained, cleared of stones, protected from overflow, or otherwise conserved where there was little if any threat to a loss of productivity. Taking the data at face value, however, some 38 per cent of the land expected to be in crops in 1975 either needed no treatment or had been adequately treated by 1958. This certainly represents a substantial accomplishment—just how great, we cannot say for sure, since we do not know exactly what kinds of land were treated.

We shall return to the matter of lands needing treatment in Chapter 9; here we will merely point out that only 74 per cent of the cropland requiring treatment has a significant erosion problem or hazard, and for some of these lands erosion is not the chief problem. For the approximately 200 million acres of cropland requiring treatment, where erosion was at least one of the serious problems, there is presumably a major problem of maintaining soil productivity; for the other 70 million acres needing treatment, it may be more a matter of building additional productivity than of maintaining what now exists. Part, possibly all, of the cropland not yet adequately treated has had

TABLE 15. Cropland adequately treated prior to 1958 in relation to the cropland area requiring treatment   (Thousands of acres)

State and region	Acreage of cropland in 1958			Cropland adequately treated prior to 1958		
	Classes II, III	Class IV	Classes II, III, IV	Acreage	Per cent in classes II, III	Per cent in classes II, III, IV
Maine	1,075	106	1,181	376	35	32
New Hampshire	245	45	290	97	40	33
Vermont	537	175	712	251	47	35
Massachusetts	263	43	306	142	54	46
Rhode Island	38	4	42	11	29	26
Connecticut	281	44	325	113	40	35
New York	5,307	934	6,241	2,317	44	37
New Jersey	820	54	874	267	33	31
Pennsylvania	5,056	1,198	6,254	1,681	33	27
Delaware	456	3	459	166	36	36
Maryland	1,700	126	1,826	558	33	31
Northeast	15,776	2,730	18,506	5,980	38	32
Michigan	9,067	1,014	10,081	2,329	26	23
Wisconsin	9,216	1,982	11,198	4,175	45	37
Minnesota	18,160	1,850	20,010	6,086	34	30
Lake states	36,443	4,847	41,290	12,592	35	30
Ohio	11,648	439	12,086	3,635	31	30
Indiana	12,241	1,032	13,272	5,045	41	38
Illinois	20,874	1,189	22,063	10,001	48	45
Iowa	21,017	1,398	22,414	8,126	39	36
Missouri	12,349	2,274	14,623	4,714	38	32
Corn belt	78,130	6,329	84,459	31,520	40	37
North Dakota	24,577	552	25,128	8,746	36	35
South Dakota	13,016	2,309	15,325	6,857	53	45
Nebraska	15,193	2,562	17,755	5,298	35	30
Kansas	22,583	2,598	25,181	7,332	32	29
Northern Plains	75,366	8,021	83,387	28,231	37	34
Virginia	3,135	547	3,682	1,242	40	34
West Virginia	871	270	1,142	450	52	39
North Carolina	6,199	853	7,052	1,837	30	26
Kentucky	4,490	795	5,285	1,871	42	35
Tennessee	5,265	1,070	6,335	2,133	41	34
Appalachian states	19,957	3,536	23,493	7,533	38	32
South Carolina	3,712	255	3,967	1,283	35	32
Georgia	5,890	535	6,425	1,754	30	27
Florida	2,233	540	2,773	629	28	23
Alabama	4,429	620	5,050	1,565	35	31
Southeast	16,263	1,952	18,215	5,233	32	29

TABLE 15.—(continued)

(Thousands of Acres)

State and region	Acreage of cropland in 1958			Cropland adequately treated prior to 1958		
	Classes II, III	Class IV	Classes II, III, IV	Acreage	Per cent in classes II, III	Per cent in classes II, III, IV
Mississippi	5,354	845	6,199	1,906	36	31
Arkansas	6,837	232	7,069	2,161	32	31
Louisiana	4,461	126	4,587	1,465	33	32
Delta states	16,650	1,203	17,853	5,532	33	31
Oklahoma	9,507	1,972	11,479	4,040	42	35
Texas	34,184	3,585	37,768	11,652	34	31
Southern Plains	43,690	5,557	49,247	15,692	36	32
Montana	11,067	1,982	13,049	7,126	64	55
Idaho	4,241	910	5,151	1,588	37	31
Wyoming	868	1,156	2,024	878	101	43
Colorado	5,231	4,629	9,860	3,664	70	37
New Mexico	1,477	511	1,988	744	50	37
Arizona	863	124	987	200	23	20
Utah	1,528	376	1,904	833	55	44
Nevada	543	149	692	270	50	39
Mountain states	25,818	9,836	35,654	15,303	59	43
Washington	6,056	1,507	7,563	2,302	38	30
Oregon	3,976	1,098	5,074	1,911	48	38
California	7,567	2,293	9,860	4,100	54	42
Pacific states	17,598	4,898	22,496	8,312	47	37
48 contiguous states	345,693	48,910	394,602	135,928	39	34
Alaska	20	3	22	29	145	131
Hawaii	180	81	261	123	68	47
All states	345,893	48,993	394,886	136,080	39	34
Puerto Rico	237	147	384	241	102	63
Virgin Islands	5	1	6	4	80	67
Total	346,135	49,141	395,276	136,325		

SOURCE: From data in *Basic Statistics of the National Inventory of Soil and Water Conservation Needs*, Statistical Bulletin 317, U.S. Department of Agriculture, August 1962.

some conservation treatment in the past; part, but an unknown area, may require relatively little further treatment to reach the "adequate" stage.

When it came to estimating the conservation needs on pasture-range and forest-woodland areas, consideration was given to the plant cover, rather than to the soil as such. The data thus are not available in terms of the area under each class of land or in terms of the soil deficiencies, as is the case with cropland.

Excluding the federally owned pasture and range land not covered by the CNI, approximately 27 per cent of the land expected to be used for this purpose in 1975 in the forty-eight contiguous states did not need treatment in 1958. (See Table 16.) The CNI instructions refer to

TABLE 16. Pasture-range and forest-woodland acreage not requiring conservation treatment in 1958   (Thousands of acres)

State and region	Pasture-range			Forest-woodland		
		Not needing treatment			Not needing treatment	
	Total area	Area	Per cent of area	Total area	Area	Per cent of area
Maine	163	54	33	17,383	9,880	57
New Hampshire	56	9	16	4,337	1,964	45
Vermont	412	133	32	4,146	2,913	70
Massachusetts	159	51	32	3,243	1,493	46
Rhode Island	24	5	21	417	167	40
Connecticut	163	48	29	1,915	1,116	58
New York	3,017	1,225	41	16,367	11,158	68
New Jersey	185	40	22	2,080	1,110	53
Pennsylvania	2,340	776	32	14,839	10,501	71
Delaware	28	5	18	391	127	32
Maryland	654	158	24	2,450	1,141	47
Northeast	7,201	2,504	35	67,568	41,570	62
Michigan	1,583	432	27	16,184	5,802	36
Wisconsin	2,792	997	36	13,945	5,525	40
Minnesota	3,158	668	21	16,075	8,069	50
Lake states	7,533	2,097	28	46,204	19,396	42
Ohio	2,835	645	23	5,158	2,101	41
Indiana	1,947	418	21	3,482	1,713	49
Illinois	2,938	1,098	37	3,633	1,045	29
Iowa	4,054	1,216	30	1,768	405	23
Missouri	8,627	3,019	35	12,728	1,799	14
Corn Belt	20,401	6,396	31	26,769	7,063	26
North Dakota	14,523	5,775	40	797	546	69
South Dakota	25,904	9,428	36	648	409	63
Nebraska	25,369	14,760	58	1,006	843	84
Kansas	19,629	5,776	29	1,156	833	72
Northern Plains	85,425	35,739	42	3,607	2,631	73
Virginia	3,752	1,466	39	13,826	8,017	58
West Virginia	2,160	854	40	10,080	5,040	50
North Carolina	1,884	531	28	17,656	8,108	46
Kentucky	5,198	1,300	25	10,832	2,537	23
Tennessee	4,028	1,060	26	11,612	5,615	48
Appalachian states	17,022	5,211	31	64,006	29,317	46

TABLE 16.—(continued) (Thousands of acres)

State and region	Pasture-range			Forest-woodland		
		Not needing treatment			Not needing treatment	
	Total area	Area	Per cent of area	Total area	Area	Per cent of area
South Carolina	1,197	303	25	11,442	5,974	52
Georgia	3,317	861	26	23,127	4,796	21
Florida	6,969	1,754	25	16,056	—	—
Alabama	4,074	1,027	25	20,263	4,237	21
Southeast	15,557	3,945	25	70,888	15,007	21
Mississippi	4,306	1,033	24	15,515	1,419	9
Arkansas	4,469	1,230	28	16,275	6,681	41
Louisiana	3,930	1,097	28	15,021	5,869	39
Delta states	12,705	3,360	26	46,811	13,969	30
Oklahoma	18,333	5,890	32	7,783	4,879	63
Texas	97,682	11,580	12	20,350	12,244	60
Southern Plains	116,015	17,470	15	28,133	17,123	61
Montana	42,717	12,444	29	6,745	3,596	53
Idaho	7,420	1,343	18	4,503	1,628	36
Wyoming	28,571	9,536	33	1,557	786	50
Colorado	22,225	5,972	27	7,436	6,322	85
New Mexico	38,330	11,965	31	7,925	6,186	78
Arizona	28,545	3,456	12	8,382	5,786	69
Utah	9,117	1,690	19	3,221	2,221	69
Nevada	7,874	1,506	19	781	633	81
Mountain states	184,799	47,912	26	40,550	27,158	67
Washington	7,174	2,106	29	13,278	8,999	68
Oregon	10,391	2,480	24	12,559	6,450	51
California	12,931	3,771	29	19,309	11,203	58
Pacific states	30,496	8,357	27	45,146	26,654	59
48 contiguous states	497,154	132,991	27	439,682	199,885	45
Alaska	62	35	56	—	—	—
Hawaii	742	135	18	1,849	787	43
All states	497,958	133,161	27	441,531	200,672	45
Puerto Rico	766	288	38	454	294	65
Virgin Islands	21	8	38	32	27	84
Total	498,745	133,457		442,017	200,993	

SOURCE: From data in *Basic Statistics of the National Inventory of Soil and Water Conservation Needs*, Statistical Bulletin 317, U.S. Department of Agriculture, August 1962.

the "area not needing treatment or not feasible to treat," whereas the publication containing the data refers only to land not requiring treatment. This percentage varies somewhat more from state to state than did the percentage of cropland not needing treatment; the spread for two-thirds of the states is 15 percentage points. No information is available as to the proportion, if any, of the area not needing treatment because it never required treatment and the proportion that had been adequately treated before 1958. There is no information as to the class of land, or the specific forage type, or the kinds of treatment which had been carried out on these lands, or as to the severity of the erosion or other problems, if any, or as to who carried out any treatments that have been made, or as to the costs incurred, or as to the role of the public agencies. It is somewhat reassuring to know that in the judgment of the soil conservationists responsible for the CNI, 27 per cent of the pasture and range land do not require conservation treatments, but it would be very helpful to know much more about these lands. There is no indication as to the extent of treatments, short of full adequacy, previously made on pasture and range lands.

For the nonfederal forests and woodlands, 45 per cent of the area expected to be in this use in 1975 in the forty-eight contiguous states was reported as not needing any conservation treatment in 1958. This proportion varied greatly between states; in order to include two-thirds of the states, one must include a spread of 35 percentage points. Florida and Alaska reported there were no forest and woodland within either state that did not need conservation treatment. On the other hand, twelve states reported that 65 per cent or more of the private forest and woodland needed no conservation treatment; these included North Dakota, Nebraska, Kansas, Colorado, New Mexico, Arizona, Utah, and Nevada—states which have either very little forest area or very little private forest land. The percentage of forest and woodland not requiring conservation treatment was low in several of the southern states. These figures seem somewhat misleading in any case, since the needed conservation programs (which we examine in Chapter 9) include control of fires, diseases and insects, and animals on a good deal of the land for which no conservation treatment is needed. Perhaps these were considered only as desirable management techniques, but they may be critical for the health and vigor of the tree growth and reproduction.

For the forest-woodland area, as for the pasture-range, there is a complete lack of the reasons why these lands do not need conservation treatment—did they never need such treatment, or have they had it?

Also, there is a lack of information as to kinds of treatment given by whom and at what cost, severity of the problems that were treated, the role of public programs, and the like. Similarly, for neither type of land were there data as to areas treated while in crop use but since shifted to grazing or forestry use; nor is there information as to the extent, if any, that the shift in use itself constituted the needed treatment for conservation. When the remaining conservation problems needing treatment on these lands are considered, we shall attempt to estimate how much of those treatments are for the preservation of the basic productive capacity of the land and how much are for increases in output within present productive capacity. In the meantime, we say only that a good deal of the proposed programs fall within the latter category.

A direct comparison between the situation in 1934, as shown by the soil erosion survey of that year, and the situation in 1958, as shown by the CNI, is impossible—at least from published data. A direct comparison would have been difficult in any case, and possibly the categories and definitions used in 1934 were unsuitable for use today; but the channeling of the CNI toward needs and treatments, rather than toward reporting of the basic factual situation, completely precludes such a comparison of the two dates. It is unfortunate that this should be the case, for a direct measurement of achievement, and of failure, during the twenty-four years between the two surveys would have been extremely valuable.

The Agricultural Conservation Program and Its Results

The second major source of information on soil conservation accomplishments over the past thirty years or more, as we have noted, is found in the reports about the Agricultural Conservation Program. To recapitulate, the ACP was started in 1936 after the original agricultural adjustment program, which sought to provide control over agricultural production, was declared unconstitutional by the Supreme Court. In order to continue a program of economic aid to farmers, emphasis was placed upon soil conservation—an objective which was not definite, but which was and is approved by nearly everyone. The original task of the ACP was to divide all crops into "soil-depleting" and "soil-conserving" crops. The soil-depleting crops were generally the row crops which were in surplus; the soil-conserving crops were mainly such close-growing crops as hay and pasture. Instead of supporting prices of farm commodities and paying farmers for taking land out of cultivation, payments were made for shifting acreage from soil-depleting to soil-conserving crops. Murray R. Benedict says in this connection, "the shift in orientation from prices to soil conservation was largely an expedient, designed to retain authorization for making payments to farmers."[4]

The original ACP was modified in the Agricultural Adjustment Act of 1938 and in subsequent amendments. Nevertheless, the basic pattern of payments to farmers for soil conservation practices has continued, while the name of the agency, the precise terms on which the payments are made available, and other details have changed considerably. Beginning in 1944, payments for shifts from soil-depleting to soil-conserving crops were no longer made; other changes in emphasis occurred in 1954, to which reference will be made later. This program, throughout its history, has been widely criticized within and outside of agriculture. One major basis of this criticism has been that ACP paid farmers to do what they should have done, or would have done, in their own economic interest. It is not our intention to discuss in detail either the virtues or the deficiencies of this program, or to present a judgment as to its social wisdom; instead, taking ACP as an accepted fact, we shall try to indicate the difference it has made to soil conservation.

Others, as well as Benedict, have suggested that the purposes of the program were deliberately confused: presented as a soil-conservation program, in reality it was largely an income-assisting program. Nevertheless, some performance or action was required from farmers in order to earn the payments made. How much of this performance or action farmers would have undertaken in the absence of the program is another story. In some states, or for some years, practices apparently had to be found to enable farmers to "earn" the funds which had been made available on a state basis by the federal appropriation. In the formula for distributing these funds among states, to be considered later, there was more emphasis on the number and size of farms than on the soil conservation problem; but the individual farmer had to have some kind of a conservation practice to justify his receipt of payments. Under these circumstances, it seems fairly certain that the program stimulated some soil conservation that otherwise would not have taken place; how much, it is very difficult to say.

---

[4] Benedict, *Farm Policies of the United States, 1790–1950* (New York: The Twentieth Century Fund, 1953), p. 351.

In Chapter 1, we distinguished among: (1) maintaining the original or basic productive capacity of the land and its vegetation, (2) building or developing new productive capacity, and (3) varying inputs of productive factors along an existing input-output curve. The ACP seems to do all three, in differing and somewhat unknown proportions. One or two illustrations help to show why these sound conceptual ideas, in practice, are hard to apply to a program such as this.

A farmer, faced with a gullying field and with severe sheet erosion, decides to construct terraces in order to prevent further deterioration of this land. The movement of earth to construct these terraces is clearly for the purpose of preventing further loss in basic productive capacity. This same purpose would stimulate the removal of stone fences which cut across the field so that terraces could not be constructed efficiently, and also the erasure of old but badly designed terraces. In addition, the planting of grass in the waterways, where the excess water arrested by the terraces would flow off, could help to reduce soil loss; and in order to get some kinds of grass to grow, it might be necessary to lime the land. All of these are capital investment in the private income-accounting sense; each has been paid for, at one time or place, as part of the ACP. Moreover, this program may result in some temporary loss of income to the farmer until a new rotation and cropping program based upon the terraced fields comes into full production; and this temporary income loss can also be considered a capital investment.

Up to this point, one could describe this hypothetical program as one of maintaining the existing basic productive capacity of the land. But, suppose that after the terracing program is finished and the new cropping program is well established, the land has greater basic productive capacity than before the treatment? This sometimes or often happens; the necessary measures that maintain productivity also increase productivity, especially if combined with somewhat more intensive and better farm management than formerly. What about the maintenance of these terraces and the necessarily associated waterways? It might be argued that the necessary costs are part of normal operating costs of the farm. But if these operations are neglected, the land may start another cycle of downgrading. Shall the necessary measures be considered productivity-maintaining, thus of a conservation nature, and hence to be paid for out of public funds for soil conservation? What about the rationale of using public funds for such practices or programs, which may have productivity-maintaining, productivity-building, and input-output features intermingled? In particular, what

about the rationale of continued payments for liming—a practice necessary to grow legumes particularly? It can be argued that such crops
help greatly in soil conservation (except on level land, where they are
not needed); but it is true that such crops also have significance for
farm income and for farm output. This latter question is not academic;
rather, it is a major issue.

A somewhat similar situation exists for a rancher with a badly depleted
range. The soil may or may not be suffering from accelerated erosion;
the original plant cover has been badly depleted but, in recent years,
the trend may have been either downward or stabilized. Such land
requires, or may need, a soil conservation program to prevent further
losses in basic productive capacity; it surely needs a program to restore
past productive capacity or even to exceed it. The most practical program may be to plow the land and reseed it. But to make reseeding
effective the rancher may have to fence the land—perhaps to cross-
fence it; develop more and better livestock water facilities; and protect
it in various ways from too much use by domestic livestock or wild
animals or rodents or from invasion of undesirable plant species. All of
this requires an investment. Again, there is an intermingling of
productivity-maintenance, productivity-building, and input variation
along existing input-output curves. Again, there is the problem of the
extent of the public interest in the investment program and in maintaining any investment. One deceptively simple answer is to say that
the public interest extends to the investment necessary to maintain
productivity, or perhaps also to the building of it, but that annual
maintenance should be a private responsibility. A failure in annual
maintenance, however, would soon bring back the need for further
investment; it may be far cheaper to maintain than to neglect and
rebuild. If the public had an interest in building, is there a public
interest in maintenance?

These problems can be put in better perspective by examining in
more detail the program for a specific year—for 1961, as an example. In
1961, total ACP payments amounted to $238.9 million; these were
shared by 1.2 million farms, or 24 per cent of all farms in that year.
These farms included 433.6 million acres of all kinds of land, or 36 per
cent of all land in farms (including 182.8 million acres of cropland,
or 39 per cent of all cropland; and including also 184.8 million acres of
pasture and range land, or 34 per cent of the total privately owned
land used for these purposes). Thus, one can say that these payments
reached from a fourth to a third of the farms and acreage. No data
are available as to the total acreage within these farms that were

involved in payments, although data show the acreage covered by payments for some practices. There is nothing in the published reports to indicate the extent to which two or more practices were carried out during this year upon the same tract of land; and it is known that much land has been included under two or more programs over the past twenty-five years.

The ACP program for 1961 was divided into a number of practices in categories and subcategories which are broadly described along with related expenditures in the following list:

Categories and subcategories

In Millions

I. Conservation practices with enduring
   benefits $177.5

   A. Establishment of permanent plant
      cover $84.2
       1) Permanent cover for erosion con-
      trol or land-use adjustment $31.9
       2) Liming materials to permit the
      use of conserving crops 34.6
       3) Increased acreage of cover in crop
      rotations for erosion control 6.8
       4) Rock or colloidal phosphate to per-
      mit the use of conserving crops 4.5
       5) Miscellaneous practices 6.4

   B. Measures primarily for improvement
      or protection cover $39.8
       1) Improvement of established cover
      for erosion control $12.7
       2) Control of competitive shrubs on
      ranges or pastures 7.0
       3) Wells for livestock water to im-
      prove grassland management 3.0
       4) Reservoirs for livestock water to
      improve grassland management 10.4
       5) Improvement of a stand of trees
      on farm land 2.2
       6) Miscellaneous practices 4.5

In Millions

C. Measures primarily for the conservation or disposal of water $53.4

   1) Establishment of sod waterways .... $ 5.0

   2) Construction of terraces to control erosion or conserve moisture 5.0

   3) Construction of permanent open drainage ditches to dispose of excess water 5.6

   4) Installation of underground drainage to dispose of excess water 12.0

   5) Reorganization of irrigation systems to control erosion and conserve water 6.0

   6) Leveling of irrigable land to control erosion and conserve irrigation water 5.3

   7) Lining irrigation ditches to prevent erosion and conserve water ..   4.2

   8) Miscellaneous practices 10.3

II. Conservation practices with benefits of limited duration .... $ 32.0

A. Measures primarily for establishing temporary protective cover $24.7

   1) Establishment of vegetative cover for winter protection from erosion $16.1

   2) Miscellaneous practices 8.6

B. Measures primarily for temporary protection from erosion .... $ 7.3

   1) Stubble mulching to control erosion and improve permeability .... $ 4.6

   2) Miscellaneous practices 2.7

III. Naval stores conservation program .... $ 0.9

IV. Supplemental (emergency) ACP .... $ 8.8

       Total of categories I through IV $219.2

Merely reading this list raises some questions about the applicability of some of the practices to soil conservation, at least in the sense of maintaining basic productive capacity of the land. There is no firm basis for evaluating them, partly for reasons suggested in the two examples given above. Nevertheless, it seems desirable to hazard some evaluation, even if it may be wrong in part. We suggest that these practices might be put into three groups:

Group I would include those practices where there is a strong probability that the chief result is to maintain the productivity of land which would otherwise deteriorate. We would put establishment of a permanent cover for erosion control or land-use adjustment, establishment of sod waterways, and construction of terraces to control erosion or conserve moisture in this group, as well as perhaps a few minor miscellaneous programs ancillary to or supporting these major programs. Their cost in 1961 was somewhat over $40 million, or more than a sixth of all the programs that year.

Group II would include those practices where there is probably a major, but not dominant, effect in favor of maintaining present basic productivity of the land; while these practices have important conservation effects, their other aspects are also important. As part of this group, we would include increased acreages of cover in crop rotations for erosion control, improvement of established cover for erosion control, various measures primarily for establishing temporary protective cover, various measures primarily for temporary protection from erosion, and the supplemental or emergency programs. Total expenditures on these practices in 1961 were about $60 million, or roughly a fourth of the total. Some, such as improvement of established cover, may increase productivity, but it is not clear that present productivity would decline in their absence if adequate management were provided for these lands. Others have essentially short-run effects.

Group III practices would be those where building of new productive capacity or increasing output seems to overshadow any effect of maintaining present productive capacity. We would include here the other listed measures—liming materials, phosphate materials, control of competitive shrubs on range land, the various programs for improved livestock water, improvement of established stands of trees, various kinds of drainage and irrigation improvements, and the naval stores program. Their cost in 1961 was nearly $100 million or about 40 per cent of the cost of this program.

The payment for limestone applications deserves special note. Proposals have been made repeatedly by the executive branch of the

federal government to discontinue these, but the political influence
of the limestone producers' trade association and others is sufficiently
great that each time they have been restored in Congress. Furthermore, all of these programs have some effects on income beyond the
borders of the individual farm—the producers of the raw materials as
well as those who handle and market the output. Limestone probably
has some conservation effect, especially on hilly land, and payments
induce the use of more than would be used in their absence. Yet it is
highly doubtful if the net conservation effect, however it might be
measured, is as great from these payments as might be obtained from
use of the same money in other ways. In recent years, roughly half of
all limestone applied has been included under payments for this
practice; in Michigan, in several years out of the past twenty-five,
virtually all of the limestone was applied under ACP payments.[5]

Melvin L. Cotner has concluded that slightly more than half of the
ACP payments have both a soil-conserving and an output-increasing
effect, and slightly less than half have only a conserving effect.[6] While
there has been some variation from year to year, the proportions have
been rather stable. More interesting is the variation he finds among
states or regions. Much less of the ACP money in the major wheat-
producing states has been used for increased output than in the major
corn and cotton states; a somewhat larger proportion of the payments
went into output-increasing measures in low-income states than in high-
income states. The average size of farms receiving ACP payments has
been considerably larger than the average of all farms in most states
and regions. The ACP program has been both a means of transferring
income from nonagricultural sectors of the total economy to agriculture, and also a means of differential income-sharing within agriculture.

In thus noting that some of these so-called conservation payments
have an output-increasing effect, we again wish to point out that
several other public programs also had this effect, and the purpose of
still other concurrent programs was to control or decrease agricultural
output. The extensive research programs of the U.S. Department of
Agriculture and the land-grant colleges were output-increasing, as
were water management programs in other federal departments. The

[5] A. Allan Schmid, "Public Assistance to Land Resources in Michigan under the
Agriculture Conservation Program," *Quarterly Bulletin,* Michigan Agricultural
Experiment Station, East Lansing, Mich., August 1961.

[6] Cotner, *The Impact of the Agricultural Conservation Program in Selected
Farm Policy Problem Areas,* Agricultural Economics Mimeo. 943, Department of
Agricultural Economics, Michigan State University, East Lansing, Mich., March
1964.

conflict between acceleration and braking actions may seem sharper when payments are made directly to farmers for these contrary purposes, but the results are not greatly different when the stimulus is less direct.

Some major changes have taken place in the ACP since it was begun in 1936 as Figure 26 illustrates. From 1936 through 1943, total payments to farmers averaged over $400 million annually; the greater part of these went to pay farmers for diverting cropland from soil-depleting to soil-conserving crops. Payments for conservation practices as such rose from $60 million to slightly more than $200 million—or half of the total annual payments to farmers—during these same years. Beginning with the 1944 program year, the diversion payments were stopped, and the conservation payments expanded somewhat. In the 1953 program year, a sharp change in the nature of payments began, with a shift away from many small payments, of which many were for practices of a temporary nature, to fewer payments for larger projects of greater long-run effectiveness. The number of participating farmers fell almost

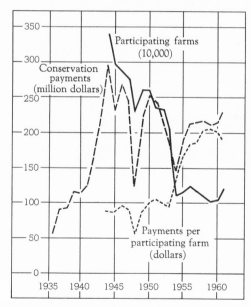

Figure 26. The Agricultural Conservation Program: total payments, number of participating farms, and average payment per participating farm, from 1936.

TABLE 17. Summary of acreage covered by selected practices under the Agricultural Conservation Program, 1936–1961

(Millions of acres)

Practices[a]	Total acreage covered in 25-year period
Measures primarily for initial establishment of permanent cover:	
Contour farming, row crops	100.8
Contour farming, close-seeded crops	38.2
Strip cropping, contour	6.0
Strip cropping, not on contour	105.1
Establishment of permanent cover	94.9
Liming (439.5 million tons)	220.0[b]
Planting trees or shrubs	3.3
Subtotal	568.3
Measures primarily for improvement or protection of cover:	
Improvement of established cover	14.9
Improving a stand of trees	2.6
Natural reseeding by deferred grazing	221.5
Livestock water, new wells (195,120 wells)	
Livestock water, springs and seeps (76,554 units)	173.9[c]
Livestock water, pipelines (44.7 million feet)	
Control of competitive plants	44.4
Contour tillage on range or pastureland	4.2
Subtotal	461.5
Measures primarily for conservation or disposal of water:	
Standard and broad-based terraces (7,460 million feet)	26.1
Spreader and diversion terraces (682 million feet)	2.4[d]
Permanent sod waterways and outlets (31.6 billion square feet)	.7[e]
Establishing permanent cover on dams, etc. (2.9 billion square feet)	.1[f]
Drainage	41.7
Storage type dams, all purposes (1.7 million dams)	5.1[g]
Erosion control structures (3.1 million structures)	3.1[h]
Leveling land for irrigation	7.5
Subtotal	86.7
Conservation practices of limited duration:	
Establishment of cover in rotation	248.9
Establishment of annual cover and green manure crops	431.8
Mulching orchards, vineyards, cropland	11.5
Weed control	19.1
Subsoiling, deep plowing, tillage to control erosion	158.3
Crop residue management	161.1
Subtotal	1,030.7
Total	2,147.2

[a] This breakdown follows the general scheme of grouping as that used in the statistical source for this table.

in half in a single year; and, for a few years, total payments declined sharply also. An irregular but modestly downward trend in total payments is evident since they hit their peak in 1944; the trend in the number of participating farms has gone down still more steeply, especially since 1953; and payments per average farmer have about doubled since 1953. While the total number of all farms participating in the ACP declined, the total number of farms in the United States also was declining, but less rapidly. There is some indication that the program shifted more toward output-increasing measures after 1954—there was a considerable increase in area of land drained, for example. In spite of these considerable changes, one can conclude that the program has been exceptionally stable—possibly too stable—considering the very great changes which have taken place in production, total number of farms, number of farm people, farm parity ratio, and other factors affecting agriculture over the past generation.

How much has been accomplished, in physical terms, by ACP? This is not an easy question to answer, but available data on the total acreage covered by the program, for selected practices since 1936 are presented in Table 17. Data have not been tabulated by the agency as to the proportion of total program costs that are included within these selected practices, but presumably the practices make up the greater part of the total program, as they did in 1961.

Using generally similar categories of practices to those given in the list of 1961 activities, Table 17 shows that measures for the initial establishment of permanent cover were carried out on 568 million acres over the whole 25-year period from 1936 through 1961. Most of the land so included was cropland. This is a somewhat misleading description of the activities, since the dominant programs, as measured in terms of total area, were liming, strip cropping, and contour farming, but this is the name given to this group of practices in the official reports. This total acreage is to be compared with 447 million acres of cropland in 1958, as shown by the CNI study. On this basis, it can be

Footnotes to Table 17 (continued).
 ᵇ Estimated on the basis of 2 tons per acre.
 ᶜ Assuming each well or other source served 640 acres and that pipelines only helped to make this possible.
 ᵈ Estimated on basis of the same proportion between length of terraces and area as with standard terraces.
 ᵉ Actual area of waterways—service area is many times greater.
 ᶠ Actual area of cover on dams—service area is many times greater.
 ᵍ Estimated on basis of 3 acres each.
 ʰ Estimated on basis of 1 acre each, including protection of outlets and inlets.
 SOURCE: *Agricultural Conservation Program-Summary by States, 1961*, Agricultural Stabilization and Conservation Service, U.S. Department of Agriculture, November 1962, Table 3, pp. 64–69.

seen that, on the average, each acre was treated one and one-fourth times during the 25-year period. Actually, some land was not treated at all, hence other land was treated more often than the average. No data are available as to how many times different tracts were treated by one or more of these practices, and as to how much land was never included in the total.

Some further light on the effect of various programs can be shed by looking at groups of practices on cropland in selected states. (See Table 18.) Some kinds of programs are essentially permanent; they need to be undertaken only once, if done adequately, and thereafter maintained. Included among these are the construction of standard and spreader terraces, establishment of permanently sodded water-ways, drainage of cropland, and leveling of land for better irrigation. The area included in the group of permanent practices in Texas over the 25-year period was 11.8 million acres; the CNI reports that 11.7 million acres of cropland in Texas had been adequately treated for conservation up to 1958. On this basis, it would seem that all the ade-quately treated land had received ACP financial assistance for at least one practice. However, this probably is not true. For one thing, the CNI data refer to 1958 only, and some of the ACP data on treated areas are for later years, from 1958 through 1961. Total crop acreage declined from 1936 to 1961, and probably some land on which practices were installed in the earlier years was shifted to other uses by 1958.

TABLE 18. Cropland area reported in permanent and semipermanent Agricultural Conservation Program practices in 1936–1961 and reported as adequately treated in 1958, selected states

(Millions of acres)

State	Area in permanent ACP practices, 1936–61[a]	Area in semiperma-nent ACP practices, 1936–61[b]	Area reported as adequately treated in 1958[c]
Texas	11.8	74	11.7
Arkansas	3.8	1.6	2.2
Oklahoma	3.0	15	4.0
Alabama	1.9	0	1.6
Ohio	1.8	0.6	3.6
Wisconsin	1.8	4.9	4.2
Georgia	1.4	0	1.8
Illinois	1.3	2.3	10.0
Florida	0.9	0	0.6

[a] Construction of standard and spreader terraces, establishment of permanently sodded waterways, drainage of cropland, and leveling of land for better irrigation.
[b] Contour farming and strip cropping primarily.
[c] As reported by the National Inventory of Soil and Water Conservation Needs.
SOURCE: From data in *Agricultural Conservation Program-Summary by States, 1961*, Agricultural Stabilization and Conservation Service, U.S. Department of Agriculture, November 1962, Table 3, pp. 64–69.

More important, it seems probable that some land was treated by still other practices, or without ACP help, in order to have been adequately treated by 1958.

The semipermanent practices on cropland covered by the ACP program—contour farming and strip cropping, particularly—have to be repeated every year to be effective. It might be argued that ACP payments should be made only to help the farmer establish such practices, after which he should continue them as part of normal farm operations. The area included in these practices over the 25-year period in Texas was 74 million acres—nearly twice the total cropland area of Texas, and more than six times the area adequately treated.

Texas was not alone in these relationships. Oklahoma, Georgia, Alabama, Florida, Arkansas, Ohio, Illinois, and Wisconsin, each show generally similar relationships. These data seem to indicate that, in many cases, the presumably permanent practices actually had to be repeated on the same land, and that the semipermanent practices probably had to be repeated several times on the same land. We have no data which show just how much land was covered once, twice, three times or more, or how much was never covered at all. It also seems clear that whatever effect the ACP has had in stimulating soil conservation is rather fully, or completely, reflected in the areas shown by CNI as fully treated.

Measures primarily for improvement or protection of cover included 461.5 million acres for the 1936–61 period, as shown in Table 17. Most of this land is pasture or range, and is to be compared with the total of 485 million acres which CNI shows was so used in 1958. Thus, on the average, this type of land apparently had been covered about once in the 25-year period by a single ACP practice. However, it seems probable that much land has been covered more than once, or by more than one practice, and some (perhaps a great deal) not at all. Again, specific information is not available, but some light can be obtained by examining specific practices in specific states. Payments for establishment of permanent cover, for natural reseeding by deferred grazing, for control of competitive plants, and for contour tillage of range lands—all reasonably permanent practices, if properly maintained—were made for 116 million acres of range land in Texas over the 25-year period. Since there were 98 million acres of range in 1958, as shown by CNI, this means that every acre of Texas range land was included more than once for this group of practices. The CNI also showed that 12 million acres of range did not require further conservation treatment in 1958. Thus, ten times as much land received some form of ACP payment as was in full conservation status by 1958. Again, Texas is not alone. The area of pasture and range land receiving improvements or protection of cover ranged from about the same area shown by CNI as adequately treated in Illinois, to three times as much in Oklahoma, to five times as much in Georgia. Either none of the land on which these practices were applied was not considered adequately treated when the CNI was made, or else several treatments were required on the same tracts, or some of each.

The conservation practices of limited duration in all states included a total of over 1 billion acres, or more than two and a quarter times the cropland acreage to which they might apply. These are admittedly used to deal with a seasonal or temporary situation, and must be repeated annually or frequently. For a number of the more important agricultural states, the acreage covered in these practices ranges from an area equaling the total cropland area to an area four or five times greater than the total cropland for the state, and from approximately three to fifteen times the area shown by CNI as adequately treated in 1958. Annual practices, such as a winter cover crop to reduce erosion from winter rains, may prevent serious soil loss during the year of application, but they must be repeated every year to be effective. The costs to the public of such practices may be relatively low, yet it seems clear that paying for them once did not lead the farmer to undertake them thereafter at his own expense.

How much of the ACP was a net gain—that is, how much remained after the twenty-five years were over? Of course, the program did not stop in 1961, but that is a convenient point for assessment. The analysis so far would suggest that the area brought to an adequate degree of conservation and maintained there was relatively small compared with the area on which payments were made. Some of the area shown by the CNI as adequately treated was surely treated without ACP help, and some may have arrived at an adequate treatment stage by means of simple practices not included in the selected list discussed above. Some land must have been treated as a result of receiving ACP payments two, three, or more times, for different or for the same practices, in order to get to an adequately treated stage; but payments must have been made on much more land without producing sufficient improvement for the land to be considered adequately treated. This does not rule out the possibility that some improvement was made as a result of such practices. In this, as in so many other aspects of the soil conservation situation, the information needed to arrive at a sound judgment is not available.

Where were the ACP funds spent, and how does this compare with the magnitude of the conservation problem in the same state? The geographical pattern of the ACP expenditure seems to have been well-established at least by 1946, and not to have changed materially since then. In fact, the actual total expenditures by states over the 25-year period 1936–61 conforms more closely to the total number of farms in each state in 1959, as shown in Figure 27, than it does to the total acreage of cropland as shown by CNI in 1958, as in Figure 28; and conforms still more closely than it does to the area of cropland in 1958 on which erosion was the dominant problem, which is shown in Figure 29. When total payments are compared with the total number of farms, there is a considerable amount of dispersion among the states, with a slight tendency for proportionately smaller total payments in the states with the most farms, and proportionately more payments in the states with the fewest farms. When total payments are compared with total cropland, the dispersion is greater; and there is also a greater tendency to smaller proportional payments in the states with the most cropland, and consequently with proportionately greater payments in the states with the least cropland. But these tendencies become most marked when total payments are compared with the area of cropland which has erosion as a dominant problem. The correlation here is much lower. These comparisons do not lead to a conclusion that ACP money is spent predominantly in relation to conservation needs.

When it comes to payments for specific practices, there is a much greater regional specialization. Some kinds of practices are appropriate in only a few states, others have wider applicability. Two-thirds of the total payments over the 25-year period for each of the following nine practices were spent in the following groups of states:

Contour farming with row crops .... Texas, Iowa
Contour farming, close seeded crops .... Texas, Oklahoma, Kansas,
                                          Nebraska
Contour strip cropping .... Wisconsin, Pennsylvania,
                                          Texas, Minnesota,
                                          South Dakota, Iowa,
                                          North Dakota, Colorado
Sodding of permanent waterways .... Kansas, Nebraska, Iowa,
                                          Illinois, Texas, Wisconsin
Building standard terraces .... Texas, Kansas, Oklahoma,
                                          Alabama, Nebraska
Mulching cropland .... Texas, Nebraska, Hawaii

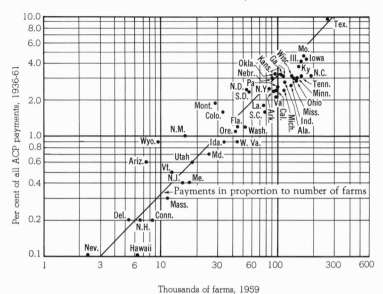

Figure 27. Total payments under the Agricultural Conservation Program, 1936–61, in relation to number of farms in 1959, by states. (Double-log chart.)

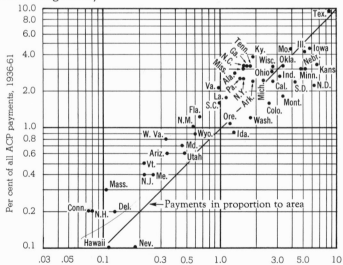

Figure 28. Total payments under the Agricultural Conservation Program, 1936–61, in relation to total area of cropland in 1958, by states. (Double-log chart.)

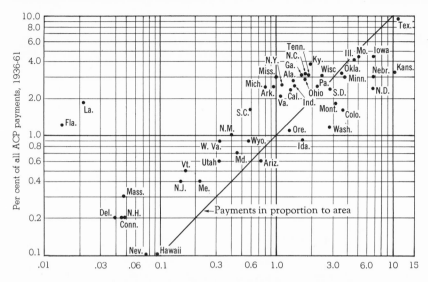

Figure 29. Total payments under the Agricultural Conservation Program, 1936–61, in relation to area needing conservation treatment on which erosion was the dominant problem in 1958, by states. (Double-log chart.)

Subsoiling, regular ....Texas, New Mexico,
                                                        California
Stubble mulching ....North Dakota,
                                                        South Dakota, Montana
Leaving stalks or stubble ....Kansas, Texas

Unlike these practices, the following were widely distributed among a rather large number of states: constructing spreader terraces, establishing permanent cover, using more annual cover, improving established cover, and using limestone (except in the West, where lime is not needed). The specific kinds of cover probably differed considerably from state to state.

The ACP program has been costly; as indicated earlier in Table 5, a total of nearly $6 billion has been spent for this program over the years. This is more than four times the expenditures for the program of the Soil Conservation Service during the same period, and it is over half of the total federal expenditures for all types of conservation. Admittedly, one basic objective of this program has been to raise farm income; its effectiveness toward this end cannot be judged without making more detailed studies in different directions than seem to be suitable for this book. The ACP has probably contributed something to soil conservation, but is impossible to demonstrate that more soil conservation had been purchased by expenditure of funds in this way than could have been obtained by use of the same funds in another way.

Soil Conservation as Shown by the
1959 Census of Agriculture

Another major source of information about progress toward soil conservation in the past few decades is the Census of Agriculture for 1959, which obtained information from farmers as to the current prevalence of certain soil conservation practices on their farms. The 1959 Census has the great virtue that it includes all farms, but it has the disadvantage that only a limited amount of information can be obtained with reasonable accuracy from such a census. Farmers can be asked about some practice, such as farming on the contour, and can usually reply accurately; but they were not asked, and many would not have known, on which class of land—using the SCS land-capability classification—the contouring was done; nor were they asked, and usually they would not have known, if the contouring constituted an "adequate treatment" of the land, in the sense that term is used in the CNI. Data obtained by the Bureau of the Census from farmers are generally summarized on a county basis, and also for farms of different sizes and types as well as county totals. There is no way to know, however, in which part of the county, or on what kinds of land, the practices were carried out, or how effective they were.

Information was obtained in the 1959 Census of Agriculture about three soil conservation practices of a fairly permanent type:

First, there was information on farms having a system of terraces on cropland and pastureland. By emphasis upon "system," presumably this would exclude single or occasional terraces of limited value, and would include only those with some reasonable claim to adequacy from the point of view of soil conservation. Construction of a system of terraces on a farm requires considerable knowledge about soils, slopes, weather hazards, and the like, either on the part of the farmer or by some technician who advises and helps him; in most cases, it also requires a rather considerable investment of cash. Once installed, if properly designed and maintained, such terraces should last indefinitely. Farmers reported systems of terraces on 32 million acres of land in 1959; an unknown proportion of this was pastureland, or land used for pasture in a rotation with cultivated crops. However, most of it probably was in crops. Permanent pastures generally do not require terracing as a means of erosion control.

Terracing on cropland and pastureland as reported in the 1959 Census is shown in Figure 30. One broad zone of terracing included the central Great Plains, from southern Nebraska southward through Texas; the other major zone was across the southeast, in the Piedmont. Texas, with 8.2 million acres, had the largest acreage of any state. In the South, Oklahoma had 3.5 million acres; Alabama, 2.3 million; and Mississippi, 1 million. In the central plains mostly, Kansas had 4.5 million acres and Nebraska, 1.9 million; and in the Corn Belt primarily, Missouri had 1.2 million acres and Iowa, 1.1 million. These eight states have about three-fourths of the national total of terraces.

The second type of census information on conservation had to do with cropland used for grain or row crops farmed on the contour. Farming on the contour is a simpler practice than terracing. While it requires a knowledge of where the contour lines are, which usually means an accurate survey, it does not require much capital investment. In many cases, new field layouts are involved, as are new soil tillage practices, which have sometimes been resisted by farmers accustomed to farming "on the square." Contour farming was carried out on 22.3 million acres in 1959. Most of this was in the western Corn Belt and southern Great Plains, with a considerable concentration in western Iowa and southeastern Nebraska. (See Figure 31.) Again, Texas had the largest acreage, 5 million, of any state; followed by Iowa with 2.7 million; Kansas with 2.6 million; Nebraska, 1.8 million; and Oklahoma, 1.6 million. No other state had contour farming on as much as 1 million acres; these five states had 61 per cent of the national total. The remaining acres were widely scattered, especially across the Southeast and up into the Middle Atlantic states.

The amount of land in strip cropping for soil erosion control was the third type of information provided by the 1959 Census. This practice is roughly comparable with contour farming so far as complexity and initial capital investment are concerned. It is especially important in the spring wheat belt of the northern Great Plains, but less important elsewhere, as shown in Figure 32. A total of 15.9 million acres was reported in 1959; of these acres, Montana had 5.6 million;

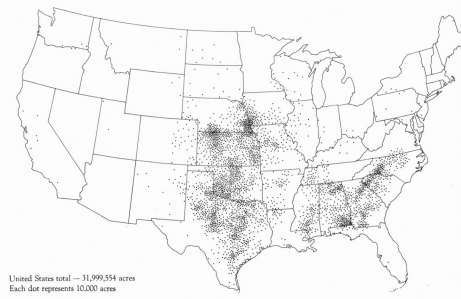

United States total — 31,999,554 acres
Each dot represents 10,000 acres

Figure 30. Cropland and pastureland with terraces, 1959 acreage. (Map from *Census of Agriculture, 1959*, Vol. II, Chap. I, p. 12.)

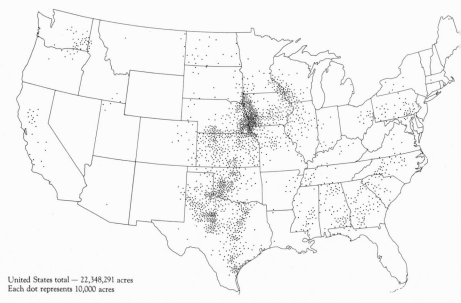

United States total — 22,348,291 acres
Each dot represents 10,000 acres

Figure 31. Cropland used for grain or row crops farmed on the contour, 1959 acreage. (Map from *Census of Agriculture, 1959*, Vol. II, Chap. I, p. 10.)

North Dakota, 2.6 million; and Nebraska, 1 million; with 58 per cent of the national total in these three states. Here, the hazard is wind erosion, and alternate strips of crop and fallow tend to reduce soil blowing. The same hazard generally exists for the limited areas found in the southern Great Plains. Relatively smaller areas are noted along the borders of Wisconsin, Minnesota, and Iowa, and in eastern Ohio and Pennsylvania; here, the principal hazard is water erosion, and this is also checked by alternate strips of various kinds of crops.

These three practices, as reported by the 1959 Census of Agriculture, have been put into effect on a total of 70 million acres; this is slightly more than one-half of the cropland area which the CNI reported as adequately treated. However, there is a very great variation among states in this relationship. For some of the Great Plains states—Texas, Kansas, Montana, Nebraska, and Oklahoma—the area of land which farmers said were in these three practices corresponds very closely with the area reported by the CNI as adequately treated. The same is true for some southern states—Alabama, Georgia, North Carolina, Mississippi, and South Carolina—although the relationship is not as close within this group. But for most of the rest of the country, the area reporting these three practices is a small, but variable, part of the total area reported as adequately treated.

Some interesting comparisons can also be made between the 1959 Census data and the ACP data on payments to farmers. The Census reported 32 million acres of terraces in 1959; ACP reported payments to help in the construction of 28.5 million terraces over the twenty-five years, 1936–61. Assuming that each source of data is reasonably accurate, this would indicate that almost 90 per cent of all terraces had some financial help from the federal government—a strong testimonial to the influence of ACP in securing adoption of this practice. The Census reported 22 million acres farmed on the contour in 1959; ACP apparently was not making many payments for this practice in recent years, but over the 25-year period paid for a total of 140 million acres. If this had been evenly distributed in each of the twenty-five years, it would have averaged nearly 6 million acres annually. However, it seems certain that payments were made at one time or another for much more than 6 million different acres; payments once made for this practice probably resulted in the practice being carried out continuously on considerable areas. This is another instance of too little information on which to base a judgment as to the continued effectiveness of the practice paid for. The 1959 Census showed 16 million acres were strip cropped in 1959; ACP paid for only 470,000 acres in 1961,

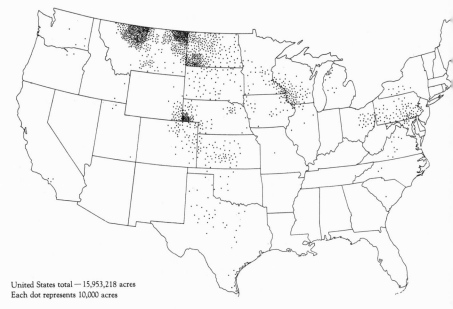

United States total — 15,953,218 acres
Each dot represents 10,000 acres

Figure 32. Land in strip crops for soil erosion control, 1959 acreage. (Map from *Census of Agriculture, 1959*, Vol. II, Chap. I, p. 11.)

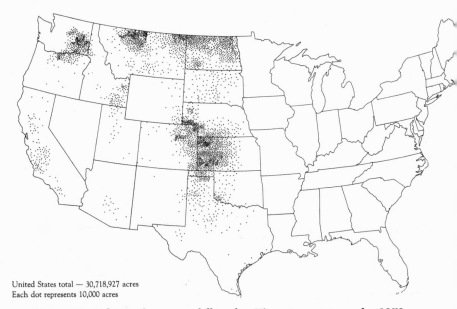

United States total — 30,718,927 acres
Each dot represents 10,000 acres

Figure 33. Cultivated summer fallow for 17 western states only, 1959 acreage. (Map from *Census of Agriculture, 1959*, Vol. II, Chap. I, p. 17.)

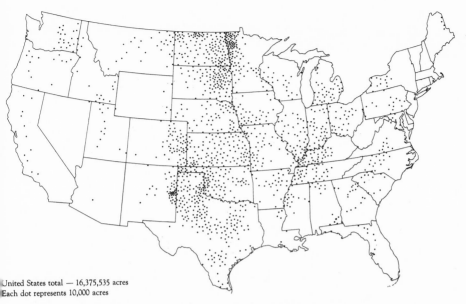

United States total — 16,375,535 acres
Each dot represents 10,000 acres

Figure 34. Land in soil-improvement crops, not harvested and not pastured, 1959 acreage. (Map from *Census of Agriculture, 1959*, Vol. II, Chap. I, p. 8.)

United States total — 9,003,704 acres
Each dot represents 10,000 acres

Figure 35. Land in cover crops, 1959 acreage. (Map from *Census of Agriculture, 1959*, Vol. II, Chap. I, p. 10.)

but over the 25-year period it has paid for a total of 111 million acres in this practice.

The 1959 Census of Agriculture also obtained information from farmers about three other soil conservation practices, which may or may not represent established permanent practices on the farms concerned or which have somewhat more limited soil conservation value than the previous group. These are cultivated summer fallow, cropland used for soil improvement crops not harvested and not pastured, and cropland in cover crops.

Cultivated summer fallow is primarily used to conserve moisture from one season into the next in order to provide a better chance for profitable crop yields. It may have soil conservation values if done properly—by "trashy" fallow or other practices which leave crop residues to protect the soil against wind erosion. But also it may actually encourage soil erosion if it leaves the soil surface bare and finely powdered—as was often the case in fallowing a generation ago before modern soil-conserving methods were developed. Farmers reported 30.7 million acres of cultivated summer fallow in 1959; this was primarily in the northern Great Plains, the central Plains, and the Pacific Northwest as Figure 33 shows. By states, Kansas, with 5.5 million acres, had the most; Montana was next, with 5.4 million; North Dakota had 5.1 million; Colorado, 2.7 million; Washington, 2.3 million; Nebraska, 2.3 million; Texas, 1.7 million; and California, 1.1 million. These eight states, which included all states with 1 million acres or more, had 85 per cent of the total area for the nation. Cultivated summer fallow is unknown east of the Mississippi, and in the first row of states west of that river. Some of this acreage reported as summer fallow may also have been included with the acreage reported for strip cropping, since both practices may be used on the same land.

The farmer using cropland for soil-improvement crops not harvested and not pastured may be participating in various government programs or may be using his land in this way without government payments. In the latter case, the objective may be fertility improvement as much as conservation. Farmers reported 16.4 million acres in this use in 1959. Figure 34 shows that this land was widely dispersed over the nation, although there was relatively more in the broad strip of territory from western Minnesota and eastern North Dakota southward to central Texas than elsewhere in the nation. But there was no concentration of this practice in a relatively few states as in the case of the other two practices.

Farmers reported 9 million acres of cropland in cover crops, which were also widely scattered as shown in Figure 35. In general, land in cover crops lies to the eastward of that reported in soil-improvement crops; the two practices seem to complement each other.

The total of this group of semipermanent practices reported by the 1959 Census is 56 million acres. When this is combined with the other group of fairly permanent practices, the total is 126 million acres, or almost as much as the 136 million acres of cropland which CNI reported as adequately treated. When the two magnitudes are compared state by state, as in Figure 36, however, only a modest relationship is evident. For such Great Plains states as Texas, Kansas, Montana, North Dakota, Nebraska, and Oklahoma, a much larger area is included in these six practices reported by the 1959 Census than the CNI judged to be adequately treated. This may be because some land is included for two or more practices, but perhaps there is more land on which these practices were applied that is still inadequately protected from erosion. For such Corn Belt states as Illinois, Iowa, Minnesota, Missouri, Indiana, Wisconsin, and Ohio, there is some relationship between these two magnitudes, but a substantially larger area was reported in CNI as adequately treated than is included in these six practices. Perhaps other means were found for treating these lands.

The 1959 Census of Agriculture contains other information which hints at soil conservation without directly measuring it. Farmers reported 23 million acres of improved pasture—not including cropland and woodland—in 1959. Presumably such land would have less erosion hazard remaining after improvement than before it. This land was found throughout the entire nation, with 64 per cent in the South and nearly a third in Texas alone. One of these hints is in the information on applications of limestone, which may help in the establishment of grass or other cover crops, and thus aid soil conservation, but is not conservation as such. According to the Census, farmers used 19 million tons of limestone on 10.2 million acres in 1959; all in the eastern half of the country, since western soils are generally not too acid without lime, and much of it was through the Corn Belt. (See Figure 37.) The only state reporting as much as a million acres was Illinois. The tonnage of limestone used in 1959 was more than double that used in 1939. The ACP shows payments for 6.1 million acres in 1960, or somewhat more than half of the total usage reported by the Census. The Census also obtained data on the use of commercial fertilizers. While

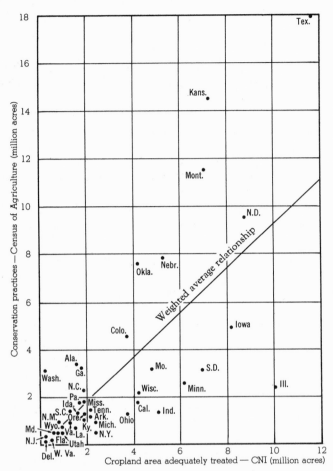

Figure 36. Relation between cropland area adequately treated and area with specified conservation practices, by states. (Note: The acreage covered in the 1959 Census of Agriculture data includes cultivated summer fallow, cropland used for soil-improvement crops not harvested and not pastured, cropland in cover crops, cropland used for grain or row crops farmed on the contour, land in strip-cropping systems for soil erosion control, and terraces on cropland and pastureland. (The CNI reference is to the 1957 National Inventory of Soil and Water Conservation Needs.)

United States total — 10,199,157 acres
Each dot represents 10,000 acres

Figure 37. Acreage on which lime and liming materials were used, 1959. (Map from *Census of Agriculture, 1959,* Vol. II, Chap. IV, p. 327.)

fertilizers are primarily increased inputs along an established input-output curve, from the conservation point of view it might be argued that heavier crops, whether row crops or hay or pasture, would protect the soil better against the hazards of erosion than would poor crops. Certainly, fertilizer might permit growing a given volume of output on a smaller area of the better soils, thus making it possible to shift some of the poorer soils to a less demanding kind of use. The Census shows that nearly two-thirds of all farms reported using some fertilizer in 1959; 19.8 million tons were used on 133 million acres. Fertilizer was applied to 98 per cent of all tobacco grown, to 75 per cent of the Irish potatoes, to 65 per cent of the corn, to 62 per cent of the cotton, and to lesser proportions of other crops.

In noting that only a portion, and often only a small portion, of all farmers apply particular soil conservation practices, these questions naturally arise: Why do so few farmers use these practices? What kinds of farms do, and what kinds do not, employ particular practices? Some modest light upon these questions can be thrown by the 1959 Census.

In nine representative types of farming areas chosen by us for analysis, on farms with less than 100 acres of cropland the use of contour farming generally was less than the average of all farms; on farms with 100 to 180 acres of cropland, use of this practice was about average; on farms of 180 to 1,000 acres of cropland it was used definitely more than average; and on farms with more than 1,000 acres of cropland its use tended to be less than average. The smallest and the biggest farms had the poorest record and the middle-sized ones, the best.

In these same nine areas, more dairy farms employed contour farming than did other kinds of farms; other livestock and cotton farms were about average; and cash-grain and general farms were below average.

When an attempt was made, by rank correlation methods, to measure the relationships for county average data, no meaningful and consistent relationships showed up between the number of conservation practices (as measured by contour farming) and such factors, among others, as the kinds of crops grown, the amount of tenancy found, and the level of family living. One can only assume that other factors intervene to cover up any relationships that might exist, or that county average data obscure as much as they reveal.

Operations of the Soil Conservation Service

The Soil Conservation Service unquestionably has been a major factor in the improvement in soil conservation that has occurred over the past thirty years or so. Its primary function has been to promote soil conservation on private lands; and it has pursued that objective with conviction—one might say, almost as a crusader. The effects of the operations of this agency since its start are reflected to a very large extent in the statistics on acreages treated and the like from the various sources already discussed, yet it seems worthwhile to look at the SCS programs separately.

One major device used by the SCS in working with private operators has been the stimulation of soil conservation districts and cooperation with them in many programs. The U.S. Department of Agriculture formulated the standard act for soil conservation districts, and urged states to adopt it. The first state soil conservation district act became effective in Indiana on March 11, 1937; by the end of 1937, acts were in operation in some twenty-two states.[7] The others came along more

---

[7] U.S. Department of Agriculture, *Agricultural Statistics, 1963*, Table 773, p. 562.

slowly; Hawaii was the last, putting its act into operation on May 19, 1947; of the forty-eight contiguous states, Connecticut was last, with its act becoming effective on July 18, 1945. The various state acts have varied considerably from the standard act; however, the various laws have been sufficient for the U.S. Department of Agriculture to sign memoranda of understanding with the districts organized under them, and to carry out co-operative programs of soil conservation.

The formation of soil conservation districts naturally had to follow the passage of the enabling acts in each state. Districts were formed quite rapidly during the forties, as shown in Figure 38. From fewer than 500 districts in mid-1940, the number climbed rapidly, to well over 2,000 only ten years later. Today, there are over 2,900 such districts—almost as many as the 3,134 counties in the forty-eight contiguous states. Some soil conservation districts are larger than

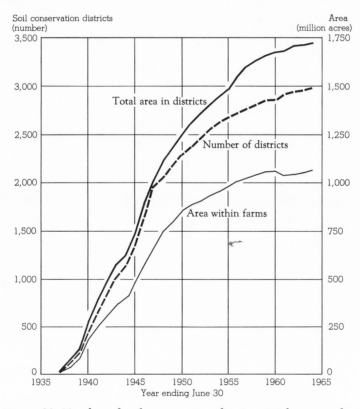

Figure 38. Number of soil conservation districts, total area, and area within farms in districts, cumulative totals, 1937–64.

counties, others are smaller, but a great many follow county lines exactly. The total area of land within districts, and the total area of land in the farms included within districts, each rose in close proportion to the increase in the number of districts.

On July 1, 1964, soil conservation districts covered most of the forty-eight contiguous states of the United States. (See Figure 39.) Their total area, in these states, as reported in the statistics, was 1,694 million acres in 1962; this is 117 per cent of the inventory acreage of the CNI. Since small amounts of CNI inventory acreage are not within districts, this is evidence that substantial areas of federal lands, excluded from the CNI inventory acreage, are included in the data on the gross area of the districts. The area reported in farms within districts in the same states was 1,039 million acres, or 72 per cent of the total CNI inventory acreage. The 1959 Census of Agriculture reported 1,124 million acres in farms in the forty-eight states; thus 92 per cent of the total area in all farms was apparently in farms located within soil conservation districts. The Soil Conservation Service reported 4.5 million farms in districts in 1960 in the forty-eight contiguous states; but this is 121 per cent of the 3.7 million farms reported by the Census of Agriculture for the fall of 1959. By 1962, SCS had revised downward the number of farms in districts to 3.6 million, or slightly less than the number reported by the Census for 1959. The number of soil conservation district co-operators in mid-1960 was reported by the SCS at 1.8 million farms, or 50 per cent of the number of farms as reported by the 1959 Census. By the same date, the SCS had made a total of 1.3 million basic farm plans in these districts; this includes 35 per cent of all farms as reported by the 1959 Census. Not all of these farm plans were in effect at any one time. In some instances, farms had been sold or rented, and the new operator either was unaware of the basic plan or unwilling to follow it for some reason. In other cases, the area of land within the farm had been changed by purchase or leasing, and the new farm was not covered entirely by a basic plan. Moreover, even basic plans made at one date may not seem entirely applicable several years later. Nevertheless, the number of farm plans is a significant measure of application of soil conservation philosophy and techniques.

The SCS program has varied somewhat from year to year, yet basically it has been rather stable. (See Appendix Table 1.) Approximately 100,000 basic conservation plans have been prepared annually; this is roughly 3 per cent of the present total number of farms. Something in the range of from 3 million to 5 million acres of newly applied

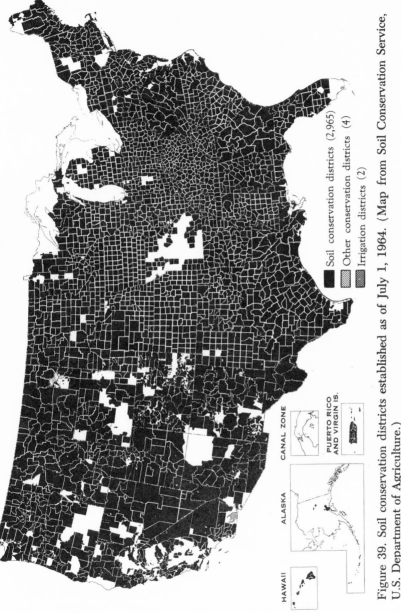

Figure 39. Soil conservation districts established as of July 1, 1964. (Map from Soil Conservation Service, U.S. Department of Agriculture.)

practices under SCS technical guidance have been carried out annually for contour farming; and about the same number of acres for cover cropping, and also for seeding ranges and pastures. From 40,000 to 50,000 miles of new terraces have been built annually, as well as from 50,000 to 75,000 new farm ponds. In addition, a smaller number of acres has had improved irrigation or been leveled for irrigation, or been drained, or had improved water application, or had trees planted annually. These various new practices cover something of the rough order of 5 per cent of the cropland each year. It is helpful to have data on the acreage of newly applied practices, as contrasted to the ACP data which simply show total areas treated annually, some of which have been treated more than once in the past. But these figures on acreage of newly treated land are not net, since some previously treated lands may have reverted to an unsatisfactory condition.

Most areas treated under SCS technical direction are also included in the areas reported by ACP, since most farmers got financial as well as technical help from the federal government. The same areas also are likely to show up in the CNI as adequately treated, if the practices were carried out properly and have not been allowed to fall into disrepair.

Wildlife conservation is a special feature of soil conservation programs. On many farms, there are small pieces of land which are not, or should not be, in crops or even in pasture. These lands generally would be placed in SCS land-capability Class VI, Class VII, or Class VIII. Often they are sides of valleys with steeply sloping land, or rock outcrops, or other physical situations where small areas are not suitable for cropping although most of the surrounding land is suitable. Sometimes there are small odd-shaped pieces of land in field corners or elsewhere. On farms in many areas of the country, windbreaks are desirable even on good land; in other situations, living fences of various species are both useful and economical.

In all of these situations, the areas not used for crops can be satisfactory homes for wildlife—especially the smaller species such as rabbits, pheasants and other game birds, and songbirds of various kinds. They require and can find shelter, water, and food in such areas; but modest special planting, such as lespedeza or multiflora rose, often can make these areas even better for wildlife, and greatly encourage additions to their numbers. In this way, opportunities can be created for more hunting, as well as more pleasure from watching the wildlife.

Many farmers set aside such areas, plant them with suitable food species, and even provide water simply for their own pleasure and interest. Such activities also generally provide more hunting oppor-

tunities for the larger public, but the farmer has no financial interest
in doing this unless he charges hunters for the privilege of hunting on
his land. The Soil Conservation Service has publicly supported the idea
of charges or fees for hunting on farms and other private land; many
sportsmen's groups and state fish and game departments have opposed
the imposition of such charges, arguing that the game is public property and that no charge should be made for its harvest. Although the
issue is rather complex, the choice may come down to paying for
hunting on private land, with relatively numerous supplies of game,
versus free hunting, with many farms closed to hunters and with a
paucity of game.

Another special feature of soil conservation has been the management of farm stockwater ponds to provide opportunities for fishing.
Under the Agricultural Conservation Program payments have been
made for the construction of 1.7 million storage type dams of all kinds
(see Table 17), and many of these are farm ponds. Some additional
ponds may have been built without federal financial aid. The size of
these ponds varies considerably but averages about 3 acres. Most of
them can provide good fishing opportunity if properly managed. The
water must be fertilized and stocked with bluegill and bass in proper
proportions, and must be fished adequately to keep fish numbers down
and the two species in proper proportion so that the remaining fish
can grow larger on the available food supply. Some colder ponds can
support trout, while the warmer ones may be stocked with catfish.
In all types of ponds, it is desirable at intervals to kill all the fish by an
application of rotenone, which does not render them inedible, and then
to restock with desired species in proper proportions.

These ponds provide fishing by means of which some food is produced, but primarily they provide fun. Since they are widely distributed throughout rural areas, they are mainly useful to farm people.
However, they could be a greater source of enjoyment for urban
people, especially people in the small and moderately sized cities.
These ponds are also a factor in waterfowl production. Successful
ponds for wildlife require adequate soil conservation on their drainage
areas, so that siltation is kept to a minimum.

Intensive Studies of Sample Areas

Analyses of totals and broad averages are informative and valid,
but they often conceal large and important shifts on small areas or on
specific soil or land situations. To understand the total and average
shifts, it is usually helpful to know more precisely what has gone on

within the small areas which go to make up the large one. For this purpose, more intensive studies of sample areas often are valuable.

A number of studies of sample areas which deal with some aspect of soil conservation have been made by the U.S. Department of Agriculture, various agricultural colleges, and others. These studies often have measured such socioeconomic factors as farm size, farm management practices, landownership, age of farm operator, and others which are, or might be, associated with current differences in soil conservation programs among the various farms in the sample area. Some studies have shown how farms could be replanned, or how crop practices could be changed; or they outline some specific soil management practices that might be adopted, or other changes, any of which would increase the conservation achievement of the farms or the area studied. Still other sample studies have considered the question of the profitability—or its lack—from soil conservation programs. All of these considerations are useful, and we draw upon them in the next two chapters.

Our present concern, however, is with the degree of soil conservation achieved over the past thirty years or so. In particular, we are interested in knowing on what types of land the soil conservation practices considered earlier in this chapter have been carried out. For this purpose, the various studies of sample areas have rather limited value. They generally do not make comparisons over historical periods, especially long ones; often they are not quantitative as to the soil conservation achievements; and they rarely relate the changes in soil conservation to land classes. Moreover, these studies are not fully comparable, which further limits their value. Despite these deficiencies for our purpose, they can provide useful insights into the changes that have occurred.

One important soil conservation area has been the southeastern part of the United States. In much of this relatively old farming region, land has been cleared, farmed, and abandoned; has grown up to trees, and then been cleared again. As Figure 16 showed, this region has severe rainfall. This hazard, with rolling topography in the Piedmont and steep slopes, together with many easily erodible soil types, have naturally led to much erosion in the past. In all of the national erosion surveys or maps (presented in Figures 6, 23, and 24), the Southeast has shown up as a region with much severe erosion. Regional statistics showing the precentage of the total area with different degrees of erosion did not show up so badly for the southeastern states as a group, because of the extensive areas of coastal plain which have no

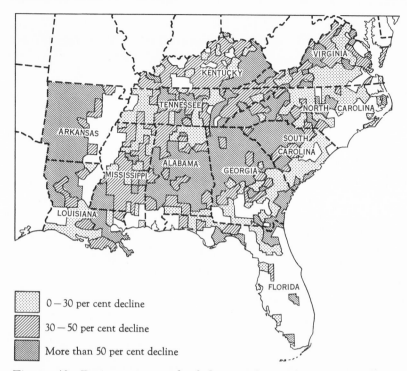

Figure 40. Decrease in cropland harvested, 1925–60. (Map from R. N. S. Harris, G. S. Tolley, and A. J. Coutou, *Cropland Reversion in the South,* North Carolina State College, 1963, p. 30.)

serious erosion problem. The erosion situations were all in the Pied-mont and mountains.

Over the past generation or longer, this general region has experi-enced major changes in land use; while these changes were caused only to a minor degree by erosion hazards, they have had a major effect upon the conservation situation.[8] A great deal of land is no longer in crops; some has been planted to pasture or to trees; and other land has gone through periods of idleness of varying length before reverting naturally to trees. In the cases where reversion has been natural, some form of ground cover came in during the first year of idleness. And under any of these circumstances, most land is rather quickly protected from further erosion.

The changes in land use have differed greatly in various parts of the Southeast (See Figure 40 and Table 19.) A study of twenty-three

[8] R. N. S. Harris, G. S. Tolley, and A. J. Coutu, *Cropland Reversion in the South,* A. E. Information Series 100, Department of Agricultural Economics, North Carolina State College, Raleigh, N.C., May 1963.

TABLE 19. Changes in cropland harvested, 1924–1959, by subregions within Southeastern United States   (Thousands of acres)

Geographic region, with numbered subregions	1924 Area	1959 Area	Change 1924–59	
			Area	Per cent
Coastal Plain:				
1. North Coastal	1,980	1,965	− 14	− .7
2. North Central Coastal	1,366	1,611	+ 245	+18.0
3. Central Coastal	5,794	5,171	− 623	−10.7
4. Georgia Flatwoods	170	79	− 91	−53.4
5. West Florida	476	350	− 126	−26.4
6. Florida	580	1,142	+ 562	+96.9
Subtotal	10,366	10,318	− 48	− 0.5
Piedmont:				
7. North	1,119	811	− 308	−27.5
8. Central	3,016	2,194	− 822	−27.3
9. Southern	6,815	2,408	− 4,408	−64.7
Subtotal	10,950	5,413	− 5,537	−50.6
Mountain states:				
10. Virginia	1,356	808	− 548	−40.4
11. Central	2,793	1,446	− 1,347	−48.2
12. Kentucky	808	294	− 514	−63.6
13. Tennessee	108	92	− 15	−14.2
Subtotal	5,065	2,640	− 2,425	−47.9
Interior Plateau:				
14. Kentucky Bluegrass	4,861	4,025	− 836	−17.2
15. Western Slope	2,402	1,325	− 1,077	−44.8
16. North Alabama	1,441	1,103	− 338	−23.5
Subtotal	8,704	6,453	− 2,251	−25.9
South Central Plain:				
17. South Mississippi and Alabama	5,658	3,029	− 2,629	−46.5
18. Gulf Coast	2,717	2,156	− 561	−20.6
19. South Mississippi	573	263	− 310	−54.1
20. West Tennessee and North Mississippi	3,813	3,011	− 802	−21.0
Subtotal	12,761	8,459	− 4,302	−33.7
21. Delta	4,363	6,631	+ 2,268	+52.0
Ozarks and Sandy Plains:				
22. Sandy Plains	5,081	1,010	− 4,071	−80.1
23. Ozarks and Central Prairies	2,469	873	− 1,596	−64.6
Subtotal	7,550	1,883	− 5,667	−75.1
Total	59,759	41,797	−17,962	−30.1

SOURCE: Adapted from Appendix Table 3, R. N. S. Harris, G. S. Tolley, and A. J. Coutu,
Cropland Reversion in the South, A. E. Information Series 100, Department of Agricultural
Economics, North Carolina State College, Raleigh, N.C., May 1963, p. 79.

subregions of the area shows that in the period from 1924 to 1959, there was a decrease of 21 million acres of cropland in twenty of the subregions, and an increase of 3 million acres of cropland in three other subregions. The 21 million acres amounted to 35 per cent of the cropland in the whole Southeast in 1924, and 39 per cent of the cropland in the subregions where the reductions occurred. Even this understates, to an unknown but probably significant extent, the shift of land out of crops, for these changes are net, and some land was surely cleared and brought into cultivation in the subregions which were experiencing net decreases. The degree of change in cropland varied greatly among subregions of the Southeast, from an increase of nearly 100 per cent over the 1924 base in Florida, to a decrease of more than 50 per cent from this base in five subregions. It is probable that the shifts were even greater, relatively, in various localities within these recognized subareas.

A number of factors have affected these shifts in land use. Basically, changing agricultural technology has favored some areas and harmed others. The small sloping and irregular fields of the Piedmont and mountain areas have not been suited to the use of modern machinery. Other areas, farther west in the South and the Southwest, have been able to grow cotton cheaper and more efficiently, and thereby have taken away from the Southeast its traditional crop. Many of the farms always were small, and provided little more than a bare existence to their operators; as off-farm employment became available, these farms were no longer operated as separate farms. Some abandoned croplands have been within farms which were still operated, at least on a part-time basis; the abandoned cropland was replaced with pasture or trees. Other abandoned croplands, especially in the past decade, have made up whole farms. The operator may continue to live in the same farm-house and work elsewhere, or he may move away from the farm. In some cases, severe gullying may have rendered land incapable of further farming, or severe sheet erosion may have lowered crop yields below a profitable point. More generally, however, erosion was probably only a contributing factor to changes in land use that were primarily determined by economic forces of the market.

What kinds of land were abandoned? Unfortunately, we can only surmise. The lands cleared and farmed were not always the best soils. One reason was that the best soils sometimes had such heavy stands of larger timber they were difficult and expensive to clear. Sloping land and poorer soils had thinner stands of timber and thus were easier to clear; if the land was abandoned, trees grew up thinly and hence were easy to clear again. A sort of cropping cycle, from trees to cultivated crops and back to trees, thus took place on some land. A major factor has been the accidents of farm boundaries. A young farmer, anxious to expand his farm business, may clear land of inferior quality because it is the best within his farm, while an older neighbor may not clear better land or may even allow better land to revert to trees.

One might expect the Conservation Needs Inventory to throw some light on this matter but, as Table 20 reveals, the data for Subregion 9, the Southern Piedmont, are disappointing. The CNI showed nearly twice as much cropland in 1958 as the Census of Agriculture reported as cropland harvested in 1959. The explanation apparently lies in the fact that the CNI included crop failure, summer fallow, idle cropland, cropland in cover crops or soil-improving crops not harvested or pastured, rotation pasture, and cropland being prepared for crops or newly seeded, as well as cropland harvested. In the Southern Pied-

TABLE 20. Area of land, according to land-capability class and use, Southern Piedmont[a]

(Thousands of acres)

Land classification and subclasses[b]		Conservation Needs Inventory 1958[c]		Census of Agriculture, Cropland harvested[d]	
		Total area	Cropland	1924	1959
Class I:	All	313	171		
Class II:	e	4,169	2,099		
	w	713	156		
	s	264	163		
Subtotal		5,146	2,418		
Class III:	e	4,583	1,256		
	w	1,182	105		
	s	994	283		
Subtotal		6,759	1,644		
Class IV:	e	4,514	613		
	w	652	14		
	s	369	66		
Subtotal		5,535	693		
Classes V–VIII:	All	9,428	202		
Total, Classes I–VIII		27,208	5,130	6,815	2,408

[a] This is Subregion 9, as defined by R. N. S. Harris, G. N. Tolley, and A. J. Coutu, *Cropland Reversion in the South*, A. E. Information Series 100, Department of Agricultural Economics, North Carolina State College, Raleigh, N.C., May 1963.

[b] The dominant hazards in the subclasses listed here are indicated as follows: e, erosion; w, excess water; s, unfavorable soil conditions in the root zone.

[c] Data from state reports in *Basic Statistics of the National Inventory of Soil and Water Conservation Needs*, Statistical Bulletin 317, U.S. Department of Agriculture, August 1962.

[d] Taken from Harris, Tolley, and Coutu, *op. cit.*

mont, the biggest item probably is idle cropland; the local workers on the CNI seem to have included nearly all the cropland abandoned since 1924, in this area at least. According to CNI, 55 per cent of the Class I land was cropland; the percentage fell rapidly to 47 per cent for Class II, 24 per cent for Class III, 18 per cent for Class IV, and 2 per cent for land in Class V through Class VIII. It is striking that so much apparently good land was not used for cropping, in the face of the comparatively large areas of poorer land used for crops. Again, the chief explanation lies probably in the accidents of land location and of farm boundaries. If only the best land were used for harvested crops, then it would not be necessary to go below Class II land in the CNI report of cropland in 1959; but this seems improbable. One may suppose that the cropland abandonment between 1924 and 1959 was proportionately heavier on the poorer classes of land, but these data supply no firm evidence. In view of these generally disappointing results for Subregion 9, calculations were not made for other subregions.

Although erosion problems were usually cured quickly on abandoned croplands by the growth of brush, trees, and weeds, in some localities, this was not true. Two sizable areas are identified in the study of *Cropland Reversion in the South*, in which the erosion problem remains serious. The report describes these areas as follows:[9]

(1) Rapid and continuous erosion on the Bluffs of North Mississippi makes establishment of natural vegetation difficult on shifting soil masses. The rolling hills overlooking the Delta are covered with deep erodible loess soils which are fragile, unstructured, porous silty loams. Cultivation of such soils, or even cattle grazing, on sloping land induces erosion which is not arrested by a heavy subsoil as the surface soil may be 40 feet deep. The soil has no self stabilizing characteristics, and consequently erosion is spectacular and highly destructive. The magnitude of the gullies raises problems beyond the capacity of farmer resources in the unprosperous region. Consequently, national resources are being used in the federally assisted Yazoo-Little Tallahatchie Watershed scheme which involves the Corps of Engineers down to the smallest farmer participant. Besides critical area stabilization by gully plugs, check dams, and reseeding to grass or trees, the emphasis has been on watershed protection by large scale afforestation and dam construction for irrigation benefits

[9] Harris, Tolley, and Coutu, *op. cit.*, pp. 61–62.

and sediment control in the Delta and in the alluvial valleys where farming in this reorganizing area has become concentrated.

(2) A second problem area is located in Georgia and South Carolina, where the Cataula and Lloyd heavy subsoil series are found. Abbeville and Fairfield counties of South Carolina in particular possess large areas of eroding abandonment cropland. The soil series are typified by a light erodible sandy topsoil overlying a sticky impermeable clay. Removal of the topsoil by erosion under cultivation has exposed the infertile clay subsoil low in nutrient and water availability. Regrowth on this unfavorable lower horizon is slow, and erosion is continuous on bare soil areas. The Louisa, Louisburg, Vance and Whitestore are similar problem soils of lesser importance. The situation is further complicated by little leaf disease prevalent on these eroded soils which cause die-off of tree seedlings.

In stating that the soil erosion problem was largely cured by this cropland abandonment, we should not leave the impression that thereafter the land was used productively. In many instances, the timber stands came back too slowly, or too thinly, for the forest to be productive; and, in some cases, various species of shrubs or brush which came in had little or no economic value. In terms of our analysis in Chapter 1, the remaining basic productivity of the soil is protected by this type of reversion. However, the basic potential productivity of the land—including its natural vegetation—is not restored by such reversion. From the viewpoint of profitable and productive land use, various forestry management programs are necessary, but the soil erosion problem is largely solved.

Another study of a sample area in the Southeast has also been made.[10] The two counties studied are within Subregion 9, the Southern Piedmont, as shown in Figure 40. The study reports on changes in crop acreage between 1953 and 1958. There was an over-all decrease in crop acreage of 29 per cent, which ranged from 12 per cent on the best land to 75 per cent on the poorest land. Unfortunately, the classes do not coincide with the SCS land-capability classes, but they are based on somewhat the same basic factors. Perhaps more striking than this differential shifting out of cropland, which would be expected, was the substantial area of lands of better class in woods, pasture, and "other open," which is presumably something of a transi-

[10] Claude C. Haren, *Conservation Farming in Selected Areas of the Southern Piedmont*, ARS 43–120, Agricultural Research Service, U.S. Department of Agriculture, August 1960.

tion stage between crops and woods; and also the substantial area of poorer lands which are still in crops. For the two counties as a whole, there was more land in the best two classes listed as other open or as pasture, than there was land of poorer qualities used for crops. However, these better open lands not used for crops probably are not parts of the same farms as are the poorer lands used for crops. Again, the accidents of ownership boundaries may affect land use as much as does the quality of the land.

What proportion of the soil erosion hazard as it existed in 1924 has been cured by these shifts in land use? A shift of 38 per cent of the 1924 cropland out of crops and largely into woods or pasture—if most of the shift were actually the poorer land with a more severe erosion problem—could well mean that as much as 60 per cent or more of the erosion hazard was solved by this change alone. It seems improbable that the land taken out was better than average, and may have been much below average. Hence the soil conservation accomplishment, through changes in land use only, probably ranged between 40 per cent and 60 per cent of the total problem as it existed in 1924 in this region. To this must be added the gains through conservation treatment of the land still in crops in 1958, as shown by the CNI. As Table 15 showed, about a third of the land still used for crops in 1958 was reported as adequately treated in this region. Based upon these combined data, therefore, one must conclude that a substantial part of the total soil conservation job as it existed in 1924 has been solved by now.

The probable soil loss per acre can be estimated by formulas which, in turn, are based upon experimental studies. One such study,[11] by Herbert Warren Grubb, gives a soil-loss equation for the Southeast as:

$$A = KRLSCP$$

where A is soil loss in tons per acre per year; K is an erositivity factor for the particular soil series, which for most southern soils runs between 0.20 and 0.50; R is an index of kinetic energy associated with rainfall distribution and storm intensity (see Figure 16), and which for the South ranges from 150 to 450, depending upon location; LS taken together represent the effect of slope length and degree of slope, in a rather complex interrelation; C is a variable reflecting the effect of different cropping patterns, and ranges from 0.001 to 0.50 in the South;

[11] Grubb, *Individual and Aggregate Benefits and Costs of Soil Conservation in the South,* unpublished doctoral dissertation, Department of Agricultural Economics, North Carolina State of the University of North Carolina at Raleigh, 1964, p. 10.

and P is the effect of physical practices for runoff control, with a value of near 0.50 when adequate physical conservation practices are used and ranging up to 1.0 when no measures are used.

In the period around 1960, soil losses on sample farms in many southern counties ranged from 6 tons to 60 tons per acre without soil conservation plans, depending upon the combination of natural factors in each locality or farm. They averaged 31 tons per acre for localities studied in Alabama; 19 tons, for North Carolina; 17 tons, for Tennessee; and 14 tons, for South Carolina. These losses presumably represent fairly well the typical farm practices in the absence of special soil conservation programs. It is entirely possible, physically, to formulate soil conservation plans for these same farms and localities which will reduce annual losses greatly, often to 5 tons or less—the objective of much soil conservation planning. With conservation plans, losses would average 9 tons annually per acre in Alabama; 7.5 tons, in North Carolina; 2.5 tons, in Tennessee; and 3 tons, in South Carolina.

Grubb does not explicitly consider actual soil losses in 1930 or some other reasonably remote past and at present; but his analysis of soil losses with different cropping systems, under different soil conservation programs, and in light of natural soil erosion hazards in various localities, does indicate changes that may have taken place. The general absence of soil conservation practices and the relatively large areas of clean-cultivated row crops in the period around 1930 must have meant a great deal of very severe soil erosion. The widespread abandonment of that cropland; a large, even if not complete, adoption of special soil conservation practices; and major shifts away from row crops toward hay and pasture—all must have combined to reduce greatly the actual soil loss since 1930. It is necessary to infer and deduce, since relevant measurements are lacking; however, we can be fairly sure of direction and have some idea of amount of change.

Another area of critical erosion hazard is western Iowa. A strip of loessal soils generally bordering the Missouri River contains about 2 million acres, most of which is rather steeply sloping land, which can and does erode easily. Originally, this land was grass-covered, with a tight sod, which protected it from serious erosion. Subsequently, the land has been plowed and corn is the chief crop—a crop growing in rows, sometimes up and down the slope and often not on the contour, which exposes the soil to water erosion. The result has been severe sheet and gully erosion (see Figure 23 and Figure 24). Because the loessal soil material is very deep—30 feet or more in many places—the sheet erosion often is not evident to persons unfamiliar with such erosion. The original dark topsoil is gone, if one but notices carefully, but the underlying soil materials are generally permeable, friable, and fertile. Gullying, however, is obvious to anyone, and some gullies are so large that it is impossible to cross them with farm machinery.

Not only was this general area recognized as one of serious erosion in the 1937 survey, but also it had been the subject of some intensive research on erosion by 1937.[12] Since that date, a number of studies of soil erosion in the area have been made by Iowa State University (formerly Iowa State College) and cooperating agencies.[13] In most of these studies, soil losses have been estimated on a sample of carefully selected farms, by means of a soil-loss formula which takes into account soil type, slope, cropping practices, soil conservation practices, and other factors—a formula which has been developed as a result of field research on the subject, and is similar to the one quoted above for the Southeast.

The average soil loss per acre annually from sample farms in the Iowa State studies was 21 tons in 1949, 20 tons in 1952, and 14 tons in 1957. The soils specialists working on this problem estimate that an annual loss of 5 tons per acre is the maximum which will keep soil-deteriorating processes reasonably in balance with soil-generating processes in this area. On this basis, roughly half of the excess soil loss, above the standard, in 1949 had been remedied by 1957. Progress was variable among farms; some were not aware of the erosion problem, or thought it impossible to do anything effective about it, and hence made little or no progress in meeting it. Soil losses in 1957 were less than in 1949 on somewhat less than half of all farms, but only about 15 per cent in 1957 were down to the limit specified by the soils specialists.

Soil losses per acre in this area of Iowa depend primarily upon what kind of crop rotation and soil conservation practices are followed. Among the cropping factors, the relative length of time in meadow—as contrasted with the time in corn or other row crops—is usually critical. Among the soil conservation practices, terraces and farming on the contour are two of the more important. On one sample farm, fairly

---

[12] G. W. Musgrave and R. A. Norton, *Soil and Water Conservation Investigations at the Soil Conservation Experiment Station, Missouri Valley Loess Region, Clarinda, Iowa, Progress Report 1931–35,* Technical Bulletin 558, U.S. Department of Agriculture, February 1937.

[13] See R. Burnell Held, Melvin G. Blase, and John F. Timmons, *Soil Erosion and Some Means for Its Control,* Special Report 29, Agriculture and Home Economics Experiment Station, Iowa State University of Science and Technology, Ames, Iowa, August 1962. This publication lists and draws upon prior studies in this area.

typical of those on the steeper slopes, the soil loss could be kept under 5 tons per acre annually if the crop rotation were corn-oats-meadow-meadow-meadow-meadow, and if the land were terraced, farmed on the contour, and fertilized. With the same rotation, but with no special soil conservation practices, annual losses on this sample farm would be 30 tons to 40 tons per acre. And on this land, with no conservation measures, and a rotation of corn-corn-oats, annual losses would run from 200 tons to 300 tons or more per acre. On the nearly flat alluvial bottom lands of this area—which are rather limited in extent—neither cropping nor conservation practices would make as much difference, because the erosion hazard is so much less.

Another sample area for which a soil conservation study has been made is southwestern Wisconsin.[14] This study states of the area: "Soil losses have been severe and intensive soil conservation practices are needed on most farms." But no quantitative measures of the extent of erosion or of accomplishments in its control are presented. Soils are mostly loessal or sandy, rather shallow to sandstone or other materials unsuitable for cropping. Thus, soil losses here are more serious than in the Iowa area just described, where at least there is a considerable depth of good soil material, even after erosion losses. The 1937 conservation survey (Figure 23) described the southwestern third to half of Wisconsin as moderate sheet and gulley erosion. The particular area included in this sample study covers nearly 3 million acres which are near the southwestern corner of the state. These lands were settled about a hundred years ago and have always been subject to some erosion; however, the past generation of farmers has been more conscious of the erosion hazard and more willing to do something about it. One major reason why it is easier to control erosion is that the area is part of the dairy type of farming region of the United States; pasture and hay do very well in it, corn less so, and various economic forces favor a larger proportion of close-growing crops, hence a smaller proportion of row crops, in this area than in the sample Iowa area.

Although this study of southwestern Wisconsin does not provide any quantitative measure of the extent of soil erosion or of the effectiveness of erosion-control measures, it discusses the nature of conservation plans for erosion control developed by SCS technicians and the percentage of those plans which are fulfilled. Conservation plans called

[14] P. O. Anderson, P. E. McNall, and Buis T. Inman, *Progress in Application of Soil-Conserving Practices, Southwestern Wisconsin*, ARS 43–44, Agricultural Research Service, U.S. Department of Agriculture, April 1957.

for about 60 per cent of the cropland in meadow each year; contour strip cropping on more than 75 per cent of the cropland; terraces and water diversion where needed, depending upon topography; pasture renovation and woods protection; and the use of about 3.5 tons of fertilizer and about 25 tons of limestone annually on farms averaging roughly 155 acres of total land and 70 acres of cropland. A rating system was devised to measure progress toward achievement of these plans; about one-third of the farms had scores of 92, another third had scores of 78, and the poorest third had scores of 55. Unfortunately, no data are presented as to what these scores actually mean in terms of conservation achievement, measured in terms of tons of soil loss per acre annually or in any other quantitative and objective way. The study did present some interesting analysis of the reasons for differences in the degree of farmer adoption of soil conservation plans and as to their profitability (which we will consider in Chapter 10).

Another study of a sample area deals with northwestern Wisconsin.[15] Although the report states that "erosion control is a problem throughout most of the area," and although descriptions are presented of the various soil types and their erosion hazards, no quantitative information is presented as to the extent of past and present erosion. The report is primarily concerned with an economic analysis of the effects of improved crop rotations and practices and improved livestock practices on farm income. "The general criterion used was that of the most intensive grain rotation permissible in light of existing erosion hazards," but continuous corn was disallowed for unexplained reasons in spite of a statement that it would be feasible on several soils when accompanied by special management practices. No specific information is presented as to the effect of these recommended cropping systems on soil losses per acre. The recommended systems show much higher incomes than do present farm systems, but this appears to be much more a matter of a much higher level of farm management than it does of soil conservation as such. This study is another example of the unfortunate tendency to accept soil conservation goals, even though nonquantitative, as desirable if not absolute, and without analysis in quantitative terms of their effects.

Another sample area studied is northeastern Kansas;[16] and, like

[15] John R. Schmidt, and R. A. Christiansen, *Potential Crops and Livestock Production and Net Farm Income on Dominant Soils in Northwest Wisconsin*, Research Bulletin 219, University of Wisconsin, Madison, Wis., May 1960.
[16] Charles C. Micheel and Charles W. Nauheim, *Economics of Soil Conservation, Northeastern Kansas*, Agricultural Economics Report 101, Kansas State University, Manhattan, Kan., December 1961.

western Iowa, it also has loessal soils and rolling topography. Although the report refers to erosion on the typical soils as "major problems," no specific data are given as to past or present soil losses in terms of tons per acre annually; and the estimated losses under different rotations range upward only to less than 14 tons per acre annually, which in the Iowa sample study would be considered quite a moderate soil loss—thus illustrating again the imprecision arising from use of descriptive terms rather than precise quantitative measurements. In this area, as in others, an annual soil loss of 5 tons per acre is accepted as permissible, and conservation plans adequate to achieve it are proposed. Estimated soil losses annually per acre run from an extreme of 13.6 tons, for a rotation of corn-corn-corn-oats-wheat and no other conservation practices, to losses of less than 1 ton per acre with terraces, fertilizer, and various rotations, including less corn at the first end of the scale. Part of the past erosion loss has been in the form of gullies so large that land has been taken out of cultivation; these range from 2 per cent to 7 per cent of the total area, depending upon the type of soil and slope.

A detailed analysis of the economics of various cropping systems has been made for the Palouse area of Washington and Idaho, which applies also to similar areas in Oregon.[17] Because of its concern with the profitability of various systems of farming and their effect upon soil conservation, we shall return to this study in Chapter 10. For our present purpose of measuring accomplishments in soil conservation over the past several years, its value is more limited. The Palouse soils are wind-blown loess, of great depth. The area is unique in appearance because of its unusual rolling appearance—it has been described as a rolling sea suddenly frozen in position. The winds apparently brought the materials from the Southwest; the southern and western slopes are relatively more gradual than the northern and eastern slopes. But the emphasis must be on the "relatively" in this statement—the average slope of the area is about 13 per cent and some land with a slope of 45 per cent is cultivated—land so steep that special farm machinery has been necessary because ordinary tractors and other machines tip over. The depth of topsoil with significant organic

---

[17] Walter W. Pawson, *et al.*, *Economics of Cropping Systems and Soil Conservation in the Palouse*, Bulletin 2, published cooperatively by the Agricultural Experiment Stations of Idaho, Oregon, and Washington (Moscow, Idaho; Corvallis, Ore.; and Pullman, Wash.), and the Agricultural Research Service, U.S. Department of Agriculture (Washington, D.C.), August 1961.

content apparently has always varied—not only according to slope but also according to position, varying from flatter areas between hills, up the slopes, and on top. The area, originally covered with dense grass cover, has been cultivated for about seventy-five years. Wheat is well-adapted to this area and yields are high. Systems of farming vary from continuous wheat, to wheat and fallow, to wheat and peas, to wheat with various amounts and kinds of legumes either cut for hay or used as green manure. Soil losses from the steeper slopes are quite evident at some seasons of the year, but as a result of the moderate rainfall and lack of strongly developed drainage systems, the total soil movement out of each locality has been quite limited.

Soil losses depend, in part, upon slope and in part upon crop rotation and soil management practices. For example, with peas planted in grain stubble, the soil loss will be less than 1 ton per acre annually; with wheat following peas, it will be nearly 8 tons; with wheat following clean cultivation, it will be more than 16 tons, but, in this case, the soil loss may not exceed 12 tons in the flatter areas while it reaches up to nearly 30 tons for the steep slopes just under the crest of the hills.

Although there has long been concern over erosion in the Palouse area, the report says, "Data presented in this report relating to loss of soil by erosion indicate that deterioration of land in the Palouse wheat-pea area through erosion is very slow."[18] It goes on to state: "With heavy application of nitrogen fertilizer and with the use of crop residues for soil conservation, the productivity of most classes of land in the Palouse would not likely be significantly impaired under most systems of annual cropping."

We close this section in which various studies of sample areas have been reviewed with considerable disappointment. A number of studies have been made but, in general, they are neither quantitative nor objective in their presentation of data about present soil erosion; they nearly all lack any real historical perspective, especially in quantitative terms; and they tend either to advocate more soil conservation or to outline ways in which greater conservation can be achieved, without much real consideration of the effects of past and present erosion or as to the rate of progress made in coping with this problem. Although we recognize that these studies often have had other objectives than ours, we cannot but express disappointment with their approach and content.

[18] *Ibid.*, p. 79.

Progress in Forest Conservation

In focusing upon soil conservation, we cannot wholly separate the soil from its vegetative cover including, particularly, forests and grazing plants. The basic productive capacity of the soil may be protected against erosion loss by totally unproductive forms of plant cover, such as brush, chapparal, weeds, and other plants suitable for neither forestry nor grazing use. Whereas, under these circumstances, the soil necessarily is not economically productive, its capacity to produce, currently or in the future, will not deteriorate further.

Volume and kinds of vegetative cover on a site may vary from zero to the maximum which that site, with its climate and soil, will support. The productive capacity of the vegetation itself, for products which man wants, may be maintained, increased (if not already at the maximum), or decreased within the reversible limits for this vegetative type—which often are very wide. It may also be lowered below the reversible limit from which recovery is not possible; for this state, the simplest case is when all members of a species are eliminated by cutting, fire, disease, or in some other way. But often a complex ecological structure exists where some type of vegetation is within the reversible range; once this is disturbed beyond some critical point, it can be recaptured only after very long periods—hundreds of years in some cases—if at all, even though the species is still present. Many of the complex forest associations at the time of earliest settlement were like that. With some of the chief species largely or wholly cut, the original ecosystem has been destroyed even though all the chief species still exist.

A great deal of the United States—nearly all the eastern half and large areas in the West—were forested when the white man first came. There was little net annual growth in these forests; fire and decay destroyed about as much as grew each year. They have often been called storage forests; they contained immense volumes of timber, but were growing very slowly if at all. Extensive areas of this original forest have been cleared and converted to cropland; some is now in the process of reversion. In other large areas, forests were cut with no thought of forest regeneration, for it was assumed—wrongly, as it turned out—that they too would be converted to farmland. The area of forested land and its volume of timber shrank steadily, so far as we can judge from the inadequate statistics, until perhaps as late as 1943–46 (see Table 21). Such data must be used with caution, partly because some of the earlier surveys may not have been very accurate,

TABLE 21. Estimates of standing timber volumes in the United States, 1903–1953

Year[a]	Estimate made by	Total volume in billion board feet[b]
1903	J. E. Deffebaugh	1,970
1908	R. S. Kellogg	2,500
1909	Bureau of Corporations—Department of Commerce	2,826
1920	Capper Report (U.S. Forest Service)	2,215
1930	Copeland Report (U.S. Forest Service)	1,668
1938	U.S. Forest Service Forest Survey—Joint Congressional Committee on Forestry	1,764
1943–46	American Forestry Association Appraisal	1,621
1945	U.S. Forest Service Re-Appraisal	1,601
1953	U.S. Forest Service—Timber Resource Review	1,968

[a] Earlier estimates were made by C. S. Sargent for the U.S. Census of 1880, by B. E. Fernow in 1895, 1897, and 1902, and by Henry Gannett of the U.S. Geological Survey in 1900. These early efforts, which paved the way for greater accuracy in later surveys, are not comparable with them and hence are not included here. The Forest Service has announced that it will publish new estimates in 1965.

[b] The total for 1909 is based on log scale; that for 1920 and subsequent entries is based on lumber tally or international scale.

SOURCE: Based on Martha A. Dietz, "A Review of the Estimates of the Sawtimber Stand in the United States, 1880–1946," *Journal of Forestry*, Vol. 45, No. 12 (1947).

but more so because there has been a major change in what was being measured. Most forest inventories have been concerned only with "commercial" timber, but what is commercial depends in part upon what is available. That is, as the larger trees of the preferred species, without serious defects, and in reasonably accessible locations were cut, and thus were no longer available, the smaller trees of the less preferred species, of shorter log lengths or with defects, and in the more remote places came to be used and valued.

A careful national survey of the forests of the nation was made in 1953 by federal, state, and private foresters.[19] It showed that the total growing stock of timber in 1953 was slightly more than in 1945; but that total live standing sawtimber had shrunk slightly. It was much more encouraging to learn that in 1953 the annual growth of all timber was up 14 per cent from growth in 1944; the annual growth of live sawtimber was up 9 per cent; and the total growth slightly exceeded the total cut plus other drains on the forests. These relationships were more favorable than in previous reasonably accurate surveys, which had nearly always shown a shrinking volume of growth and a cut more than equal to growth. At the same time, these relatively favorable

[19] *Timber Resources for America's Future*, Forest Resource Report 14, Forest Service, U.S. Department of Agriculture, January 1958.

national trends masked some serious imbalances. Regionally, growth was relatively greatest in the East and South, where forests have been most degraded over the decades; this is excellent. However, growth was relatively greatest in hardwoods, which were not in active demand, and less in softwoods, which were more in demand; and it was also greatest in relatively poor species and low-grade trees rather than in better ones.

The relatively better forestry situation in 1953 reflects many factors. For one thing, fire protection had steadily become more effective, so that fire losses were lower in spite of greater potential danger arising from the use of the woods by more people. Another major factor was that the amount of timber cut, although fluctuating considerably from year to year, has not trended upward for many years. The fact that the end of the mature old-growth (virgin) sawtimber stands is now in sight, though not imminent, has been at least partially responsible for major rises in prices of stumpage in the past two decades. These rises are of the order of three to six times, even when compared with the average of all other prices. These rising prices of stumpage have undoubtedly made forest owners more appreciative of the value of timber growth, and have spurred greater output. There are no comprehensive data about the national forestry situation since 1953. However, such surveys as have been made seem clearly to indicate further major improvement since that date,[20] although there are still unsatisfactory aspects of the present forest situation.

Progress in Range Conservation

Extensive areas of the United States are used to graze domestic livestock on native forage plants. Some of these lands are federally owned (and these will be considered separately in the next section of this chapter) but large areas are in private and state ownership.

Range lands are generally unsuited to crop production. They are usually too dry, too rocky, too steep, too infertile; or their soils are too shallow, or other deficiencies make them unsuitable for crops. This immense area is used for grazing simply because the lands are not in demand for other more productive uses. The physical output of grazing lands, as measured in pounds of air dried forage per acre, is low—on the order of 100 pounds to 300 pounds annually. Such a low

[20] The Forest Service is working on an updated review of timber resources which it expects to publish in 1965.

yield per acre will pay off only if domestic grazing animals perform the harvest operation. Despite the low physical output per acre, however, economic output per man is high—one of the highest for any major type of agriculture in the United States—because the typical rancher uses large areas of land with such low physical productivity.

The Reconnaissance Erosion Survey undertaken in 1934 called attention to the generally bad conditions on the range lands of the West. That survey—done in great haste, when time and skilled manpower and previously available information were seriously lacking—did not permit a detailed consideration of this large area. A great deal of it, as we saw in Table 13, was listed as "mesas, canyons, badlands, rough mountain land." The emphasis in that survey was upon soil erosion, not upon the condition of the range plants as such. But it was obvious that the surveyors felt that this large region generally was in poor shape and that its trend was downward.

In 1936, the Forest Service prepared an extensive report which dealt directly with the condition of the range, particularly of the plants as well as of the soil.[21] According to that report, the depletion of the various forage types, measured from their original, or "virgin", condition when white man first began to graze domestic livestock on them, ranged from 20 per cent to 70 per cent; depletion varied according to landownership as well, but averaged 51 per cent for all privately owned range. The basic productive condition of most of the range, according to this Forest Service report, was still on a downward trend. The report showed that there were 17.3 million animal units of livestock on the range in 1935, when productive capacity would support only 10.8 million animal units; and it unequivocally recommended a drastic reduction in the number of livestock as the only possible solution to the overstocked and deteriorating ranges. The report further related poor range condition to soil erosion and to silt movement by streams, and its conclusions as to erosion for different kinds of land closely paralleled its conclusions about range condition.

The tone of the report was widely resented, as being patronizing and insulting, and stirred up much adverse comment, not only among livestockmen but among agricultural specialists as well. In retrospect, it seems clear that the Forest Service was unduly influenced by the bad range conditions resulting from the extreme drought of 1934 and also from several years of below-average precipitation in many western areas. Certainly there were many unsatisfactory situations on the

[21] *The Western Range*, published as S. Doc. 199, 74 Cong., 2 sess. (1936).

range, as sober observers then realized and as are more clearly evident in retrospect; but it now seems clear that the Forest Service report exaggerated greatly, even though with sincere conviction.

Whereas no fully comprehensive survey of range lands has been made since the 1936 report on *The Western Range*, the Conservation Needs Inventory dealt with the situation on private and state lands in 1958 (as shown in Table 16), and numerous surveys have dealt with smaller areas, both public and private. The CNI shows 27 per cent of the private and state range lands, or 133 million acres, "not needing treatment"; perhaps this can be compared with the 50 million acres of privately owned range that the Forest Service in 1936 classed as in "reasonably good condition." If so, it represents a major improvement in a period of twenty-two years. As we have shown, extensive areas of private range have been treated in various ways under the Agricultural Conservation Program. Some of this perhaps needs no further treatment, and some is probably improved, but not to this point. Basic to the physical improvement of the range, and more important, have been the great increases in technical knowledge about range management and in better range practices by operators. In the years since 1935, a professional range management society has been formed, which publishes a good professional journal, and to which many ranchers now belong. Today's generation of ranchers is infinitely better informed about range plants and their proper use than was the previous generation, just as today's generation of farmers and of lumbermen is vastly better informed about their problems than were their fathers.

One well-informed specialist in range management, who at the time worked for the SCS, has expressed his opinion of changes in range conditions as follows: "I believe most people who are widely familiar with western ranges will agree that during the past 25 years there has been marked improvement in range conditions throughout almost all of this area. The startling fact is that this improvement has taken place in the face of almost continuously increasing livestock numbers, instead of the reductions once thought necessary."[22] It is perhaps significant that this statement expresses a conviction or a consensus for which no supporting evidence on a national scale is available. There certainly seems to be general agreement on the conclusion, and scattered pieces

[22] F. G. Renner, "The Future of Our Range Resources," *Journal of Range Management*, Vol. 7, No. 2 (March 1954).

of evidence are available, but no comprehensive data exist. However, as with forests, we do not wish to leave the impression that all is perfect on the range; the conservation job there is still far from done.

Soil Conservation Achievements on Federal Land

The Conservation Needs Inventory, the data from the Agricultural Conservation Program, the Census of Agriculture, and the various studies of sample areas so far discussed in this chapter do not apply to federal lands. In the forty-eight contiguous states, the CNI reported 1,448 million acres and excluded 396 million acres of federal land, as well as 58 million acres of built-up areas and water surfaces. This excluded federal area is virtually all of the federally owned land outside of built-up areas which is not in cropland. This federal land is spread among a rather large number of agencies, but the Bureau of Land Management with 179 million acres, the Forest Service with 161 million acres, and the National Park Service with 14 million acres account for by far the greater part of it except for military lands. A brief review of soil conservation accomplishments on these three major types of federal land seems desirable in order to round out the picture.

Extensive areas of this federally owned land in the form of national forests, national parks, grazing districts, and other kinds of areas are under positive management. Virtually all of this land was once open to distribution or alienation to private ownership, but for one reason or another was not taken up during the period when it was open to such disposal; and virtually all of it went through a period of unrestrained use by the public, prior to its being put under positive management. The chief form of use by private enterprisers was for grazing, but there was some lumbering and mining; generally speaking, such unrestrained use was either illegal or lacked positive legal sanction. The intensity of grazing varied greatly, from severe overgrazing in many areas to almost no use in the more remote areas. The individuals using the federal lands for grazing often underestimated the serious exploitive effects of their actions, but in any case they lacked definite control over the land and any grass their livestock did not get was likely to be taken by another rancher's livestock. Nearly all of what is now in national forests had been brought under some form of positive control by 1910, most by 1905; nearly all of the present grazing districts had been put into this category by 1938; the national parks and similar areas were put under some form of federal management at varying dates, but mostly prior to 1920. During the thirties, there was extensive purchase of submarginal farmlands, which are now designated as the national grasslands.

The nature of the management and conservation job on these various federal lands has changed greatly over the years. In the beginning, effective fire control was a major issue; also, livestock grazing and trespass were major problems. In total, however, the level of use was relatively low and the emphasis was largely upon protection of the basic productive capacity. During the thirties, relatively large amounts of labor were available through the Civilian Conservation Corps, Works Progress Administration, and other public works programs, and numerous conservation and development programs were undertaken on all kinds of federal land. The Forest Service has reduced livestock grazing on national forests in the western states (where most of the forests are found) by two-thirds from the 1918 peak.[23]

Since World War II, demand for various activities on all kinds of federal land has risen greatly. In particular, recreation use has advanced steadily, at roughly 10 per cent annually, thus doubling in about each eight-year period. The harvests of timber on national forests have risen greatly, from 3 billion board feet at the end of the war to 11 billion today. Although use for grazing has not risen, because the federal lands have long been fully stocked, the potential demand for grazing has probably increased materially. The value of water flowing off the federal lands has also risen, although the volume has remained the same. The greater use, especially the greater number of people on the federal lands and the travel that this necessarily entails, has put new pressures for greater conservation efforts.

In roughly the past twenty years, the different federal agencies have launched larger programs that are more directly concerned with conservation. The Forest Service has planted or seeded over 3 million acres of forest, from 1911 to date, of which well over half has been planted in the past twenty years. It has improved 1.8 million acres of range land by revegetation methods in the past twenty years. Each year, some 120,000 acres of damaged watershed are rehabilitated by measures such as grass seeding, contour terracing, water diversion, gully plugs, and debris dams. In addition, many other measures have been undertaken for both management and conservation purposes—fences to

[23] Marion Clawson and Burnell Held, *The Federal Lands: Their Use and Management* (Baltimore: The Johns Hopkins Press for Resources for the Future, Inc., 1957), p. 59.

control livestock better; roads to make timber stands available for harvest but also to permit better fire and insect control; improvement of timber stands by weeding, thinning, pruning, sanitation cutting, and similar measures; and so on. The Forest Service reported expenditures of over $100 million by 1956 for improvements which were directly or indirectly of a conservation nature.[24]

On the grazing districts, the Bureau of Land Management has re-seeded nearly 2 million acres of range land, has carried out brush control on nearly 1 million acres, and has engaged in many specific conservation measures such as contouring, building check and detention dams, spreading water where terrain permitted, and others. In addition, it has undertaken measures for both better management and conservation, similar to those of the Forest Service, for fences, roads, livestock water, etc. The Bureau, since World War II, has expended some $78 million on soil conservation activities, as well as smaller sums for range improvements which have often had conservation effects. The National Park Service has undertaken somewhat similar measures, except that its measures have varied to meet the heavy use of land by people rather than by livestock.

These federal programs have operated to help retain existing basic productivity of resources, to develop or build additional basic productivity, and to increase current inputs and outputs. As with private lands, it is difficult to draw a clear line between these various effects, since the same specific program often has results in all three directions. These federal programs have been generally effective. The Bureau of Land Management estimates the condition of the range in grazing districts as follows: excellent, 1 per cent; good, 15 per cent; fair, 55 per cent; poor, 25 per cent; and bad, 4 per cent. It estimates further that the condition of 14 per cent of the range is on an upward trend, 70 per cent is static, but that 16 per cent is still on a downward trend. These results, while far from wholly satisfactory, almost surely represent a marked improvement from the situation in the middle thirties—even if one does not fully accept the statements of *The Western Range* report as to the conditions at that time. The additional access roads in forests have opened up extensive areas of timber for harvest on a sustained-yield basis, as well as providing more recreation opportunity.

On the other side of the ledger, it must be recognized that further conservation efforts are still needed on the various federal lands. The demands for public land are rising steadily and greater efforts must be

[24] *Ibid.*, p. 414.

exerted to prevent deterioration of these lands in the future. One reason for the rapidly mounting demand is that use of these lands—whether for recreation, grazing, or even timber harvest—can be arranged on much more favorable financial terms than use of private lands. A private landowner, faced with greater demand than he could readily meet, would almost surely raise his prices; for various reasons, this has not been easily possible for the federal land management agencies.

Attitudes toward Soil Conservation

A factor of perhaps greater importance than all the physical achievements in soil conservation is the increasing knowledge about, and changing attitudes toward, soil conservation. Measurement of knowledge and attitudes is difficult, but not impossible; for a growing and changing field it is perhaps unusually difficult. However, the fact is that there has been no serious and continued effort to make such a measurement for the soil conservation field. Several studies do comment on this point, but they usually do so incidentally to other analyses, are limited to a single area and time, and are more on an impressionistic than on a quantitative basis.

In the period centering roughly on 1930, most farmers in the United States were unaware of the seriousness of the erosion hazard on their farms; and if aware, they were uncertain as to how to cope with it. Agricultural technicians of all kinds were almost as unsophisticated as the farmers in this respect. Few soil and crop specialists in the whole country included any mention of or showed any concern with erosion in their research, teaching, and extension work. In a great many ways, Hugh Bennett was a voice crying in the wilderness in those days. There were others who shared his concern, of course; and it is possible to find many of them, including George Washington and Thomas Jefferson, who in earlier decades had decried the wastage of soil. But all these men were exceptions; some stirred some interest but none brought about consistent and continued efforts before Bennett. Offsetting the fears of those who saw a serious erosion hazard were the many who denied its dangers and the need to undertake special programs for erosion control. For a long while, the agricultural extension services as a whole both ignored and rejected soil erosion as a subject of serious effort. Agricultural economists at that time almost never considered the erosion or conservation effects of the cropping or farm management programs they advocated. The nation, as a whole, was almost completely unaware of soil erosion.

Today, in contrast, most of the nation is "soil conservation conscious." Many people who have moderately strong feelings about the worth of soil conservation are not well-informed about it, although others are. A large proportion of the farmers still regard soil erosion as a much less serious hazard than do the SCS technicians; others may be concerned about erosion on their farms, but lack knowledge of what to do, in spite of the many programs aimed at informing and helping them; and still others think that the recommendations of the SCS technicians are impractical. As far as today's general public is concerned, it is often poorly informed about soil conservation yet almost always generally approves of it; for example, few politicians would care to run on an anti-conservation platform. Sentiment for soil conservation may be vague and poorly based, but nonetheless it is real, and it takes practical form in the approval, tacit or vocal, of appropriations of public funds for conservation activities.

These changes cannot be expressed quantitatively; we lack the data, and possibly the basic concepts, to do so. Different observers or students might appraise the changes and the present attitudes quite differently from the way we do. Yet most of the sample studies we have cited, and many others that deal with the subject less directly, comment upon changes in farmer attitudes. Differences in interpretation would be more likely to concern degrees of change, and their significance, than to deny the fact of changed attitudes themselves.

More concern by the general public, agricultural specialists and farmers with the soil conservation problem today than a generation ago is somewhat paradoxical. The seriousness of the soil erosion problem has been at least partially abated—that is the whole theme of this chapter and of the evidence we have cited. Many eroding areas have been adequately treated. Moreover, the need for concern about the future food supply is perhaps less urgent; even the most dedicated conservationist has to acknowledge the large present agricultural surpluses and to agree that scarcity is hardly imminent—though he may argue that it will be encountered some day. From the point of view of self-interest and on purely rational grounds, there was perhaps a sounder basis for "scare psychology" and "crash programs" in 1935 than there is today. In this respect, as in so many others in social and political life, there is a substantial lag between a situation and the popular response to it. The practical significance of this changed attitude toward soil conservation is difficult to measure. It seems probable that the different attitude today would call forth specific responses to particular situations which would differ from those of a generation ago—beyond this general, almost trite, statement it seems very difficult to go.

Summary

As we review the bits and pieces of evidence on the progress of soil conservation which we have put together in this unavoidably detailed chapter, it is difficult to avoid a feeling of frustration. There are simply no clear-cut, well-designed data, especially on a historical scale, which are adequate to answer the critical questions about soil conservation achievements over the past thirty years or so. The 1957 National Inventory of Soil and Water Conservation Needs comes the nearest, but it is seriously deficient. Its very name—"needs"—emphasizes its normative character. We have repeatedly noted its failure to answer the questions that must be asked if progress over these past few decades is to be appraised. The CNI and other sources provide partial indications, but not good answers. The interested American public deserves dependable objective facts on which to reach its own conclusions as to progress and "needs."

To us, however, it seems fairly clear that substantial progress has been made since about 1930. All the evidence we have been able to find points in this direction; none seems to contradict this conclusion. There almost certainly has been a trend toward improvement in the physical health of the soil and associated vegetation. This is the more significant when one considers the great increases in agricultural output which have led to huge surpluses. Over the past thirty years, American farmers with their public servants and the private industries which serve them have turned out more commodities and improved the basic productive plant at the same time.

How great has the progress been? Although the CNI, with all its shortcomings, provides perhaps the best answer, the interpretation of its results depends to a great extent upon which class or classes of land were treated. In addition to the Class I land not needing conservation treatment, about one-third of all the cropland had been adequately treated by 1958; one-fourth of the range land and nearly one-half of the forest land needed no further treatment—whether because originally needing none or because adequately treated, we cannot say. In addition, some land in need of treatment in the early thirties has shifted out of crops, and, by and large, has no soil conservation problem remaining. Moreover, in the years since the inventory was taken, further conservation should have been achieved. In addition, some land has had some conservation treatment, but is not yet considered "adequately treated." If it is assumed that the past treatments were all on the land easiest to treat, then perhaps no more than 20 per cent of the soil conservation job as it existed in the early thirties has been done by now. If it is assumed that the land with the worst soil conservation problem—and, hence, usually the land hardest to treat— has been treated adequately during the 30-year period, then perhaps as much as 50 per cent to 60 per cent of the soil conservation job as it existed in the early or mid-thirties has been accomplished. A spread from 20 per cent to 60 per cent is an unreasonably wide one, but we lack firm evidence for narrowing it. One might guess that the proportion of the job accomplished lies somewhere near the middle of this range, or a little above the percentage of the cropland adequately treated in 1958, but this must remain little more than a guess.

The soil conservation achievements of the past generation, whatever their precise magnitude, are a cause for satisfaction, but not for complacency. To stress the gains does not mean that the gains achieved were worth the cost, nor that all efforts were as efficient for accomplishing the aim of soil conservation as they might have been. But we can rather confidently base our analysis and our plans for the future on the real progress that has been made over the past several years.

Chapter 9

THE UNFINISHED
SOIL CONSERVATION JOB

How big is the remaining soil conservation job? What are the soil conservation needs of the country? It might seem that the difference between the total soil conservation job as visualized from 1930 to 1937 and the accomplishments thus far would define the job still to be done. In reality the situation is not that simple. For reasons which we shall try to make clear, the objective is neither sharply defined nor constant. Since it seems essential to look at the matter more explicitly, we attempt here a brief appraisal of the size and nature of the un-completed soil conservation job as it appears today.

Program Standards and Objectives

What are the goals or objectives of private and public programs for soil conservation? Goals might be expressed in either physical or in economic terms. In physical terms, at one extreme there might be a goal of no soil loss from any acre; but this not only would be nearly impossible to attain, but also unnecessary. A difficult, but realizable, objective would be that any soil loss should be fully offset by the formation of new soil; this rate would vary from situation to situation, but could be estimated. Physical goals have the great merit of being specific and quantitative; one can see both the target at which he is shooting and how close he comes to the mark. But they have the great disadvantage that their cost may be all out of proportion to their value; modest changes in physical goals may often reduce costs greatly. Since neither private nor public funds are unlimited, costs incurred for one set of purposes inevitably reduce the possibility of achieving other objectives.

In economic terms, one soil conservation goal might be the assurance that farmers would carry out conservation programs so far as it was profitable for them to do so. Presumably every farmer would do this, but he might be limited by lack of knowledge, by a short time horizon, by lack of credit, or in other ways. With this goal, society might help the farmer by providing more knowledge through research and extension; or by creating a more nearly perfect land market, so that the values he created by soil conservation activities would accrue to him, and thus extend his planning horizon; or by making available better credit facilities; and the like. A higher goal would be that of achieving the maximum public or social welfare possible by soil conservation programs. Presumably this would always include those activities which would be profitable to the individual farmer, plus any additional values accruing to others. For example, the soil washed off one farm often accumulates elsewhere, causing damage in the new location. There are external values from soil conservation, although their importance is sometimes exaggerated; there are relatively few external losses caused by more soil conservation on a particular farm.

What constitutes "adequate" soil conservation? If this term were limited to mean the preservation of existing basic productive capacity in the soil, then it might be possible to define "adequate" acceptably. Soil erosion is a generally harmful force or situation for any soil but, even in an accelerated form, does not necessarily represent a real menace. Soil losses differ greatly, before and after cultivation of a particular area or tract. Annual losses of 1 ton or 2 tons per acre are obviously much less serious than loss of 30 tons or 50 tons—and differences of these magnitudes often exist within local areas, depending primarily upon particular conditions on each field and upon crop practices. Moreover, the seriousness of any tonnage loss depends in large part upon the materials under the topsoil. If bedrock is close, or if the underlying material is sterile sand or gravel, or if the subsoil is stiff impervious clay, or if any one of various other unfavorable conditions exist—then much loss of topsoil is extremely serious. At the other extreme, many loessial areas have several feet of excellent soil materials; while the topsoil, with its accumulated organic material, is more productive than the subsoil, the subsoil can be developed into true topsoil in a relatively short time.

However, somewhere along the range of possible soil losses is a point for each area at which a threshold is crossed. At that point, a greater rate of soil losses over any extended period of time is likely to bring about an irreversible situation. The basic productive capacity of the soil is in danger of being permanently degraded, because physically it is extremely difficult to restore the former productive capacity, or the costs of so doing rule it out of consideration. Perhaps the word "threshold" implies a sharper line of demarcation than can be justified, but the concept is useful in establishing a limit beyond which erosion cannot continue if the land is to be productive.

A permissible rate of soil loss, which will not bring the soil to a point of irreversibility, can be established. New soil is constantly being formed from underlying materials—although at varying rates, depending upon the particular circumstances of each area. Complete elimination of soil loss often would be literally impossible and usually extremely expensive; moreover, it is unnecessary, because new soil materials are being transformed into true soil. The permissible rate of soil loss might well vary from one soil to another. A little research has been done, and much more might be done, to measure permissible soil loss more accurately under different conditions. As a general starting point, the loss of 5 tons per acre annually—a figure frequently used by the Soil Conservation Service (SCS)—is probably accurate.

When annual soil losses exceed the permissible rate, a kind of physical disequilibrium exists which, in time, will force some different kind of development. The loss can continue until the soil becomes economically and physically unproductive for crops; or some corrective measures, such as terracing or contour cultivation or use of more close-growing crops, may reduce soil loss to the permissible level; or the use of the land may be shifted drastically, to pasture or forest, where annual loss is much lower.

Political and Strategic Forces

Many conservationists, including those in the Soil Conservation Service, the Agricultural Conservation Program (ACP), and elsewhere, as we have noted in earlier chapters, have not been content with the simple goal of maintaining basic productive capacity. A number of programs have been advocated and carried out which have had the purpose, or the effect, of increasing basic productive capacity or of increasing current inputs into established production functions or both. Farmers often have not been satisfied merely to maintain present productive capacity; they have been eager to increase it, and to make greater current inputs into present productive processes, especially when they could obtain public assistance with part of the costs. Sometimes such programs have been profitable to farmers, even when they had to bear all the costs; but more frequently when part of the costs were publicly borne. Naturally enough, farmers have carried out programs which would increase their own profits. But the Soil Conservation Service has been equally interested in promoting farm programs which built new productive capacity or increased current inputs and hence outputs, as well as promoting programs to maintain existing productive capacity. That agency has had many political struggles, and it has sought to provide services desired by farmers in every part of the nation, even if this meant realignment of its original objectives.[1] Thus, drainage programs or irrigation programs, to name but two, were undertaken in some areas even though their prime purpose was more than maintenance of existing productive capacity. A political realist, noting these forces toward broader programs in the past, might well expect them in the future; and this even though the federal government might at the same time be struggling with excessive supplies of some agricultural commodities. Existence of a national objective for reduced or restricted farm output well might fail to prevent public programs whose practical effects were to increase agricultural output.

This push for programs which will result in increased agricultural output is clearly illustrated by the Small Watershed Program of the U.S. Department of Agriculture. Dams and other structures built as part of this program may protect areas from flood hazard, as well as have other values; and the sum total of the benefits may exceed the costs—although benefits often have been exaggerated and costs underestimated. But the effect of such structures is almost always to increase total agricultural output; they do not simply, or even primarily, prevent destruction of basic productivity capacity, through erosion or in other ways. Instead, they raise the basic productive capacity to new levels. The local area, including the farms so protected, benefits from such programs, even if the national benefit should be low, or zero, or even negative. And, since a substantial part of the total costs are borne by the federal government without reimbursement by the beneficiaries, there is a strong incentive for local people to seek such programs. As long as the locality's benefits exceed its costs—and local costs are usually only a small part of the total—local people have gained, regardless of the national balance sheet. Both benefited landowners and the SCS may argue accurately that the Small Watershed Program offers no more in the form of subsidies than is extended through the much

[1] Robert J. Morgan, *Governing Soil Conservation: Thirty Years of the New Decentralization* (Baltimore: The Johns Hopkins Press for Resources for the Future, Inc., in press).

larger flood protection projects constructed by the Army Corps of Engineers. Rivalry among agencies, as well as the local desire for maximum benefits, provides powerful inducements for the expansion of programs in small watersheds.

The National Inventory of Soil and Water Conservation Needs (CNI) was supposed to obtain information on what was needed in small watersheds in 1958, but the report states: "In view of developments that have increased interest in these project purposes since the Inventory was undertaken, these estimates are now considered to represent only a fraction of actual needs. . . . This part of the Inventory is considered incomplete and does not reflect the true magnitude of nonagricultural water management problems."[2] However, even this survey concluded that more than half of the total land area lies in watersheds requiring projects of this type, that "project action is required to deal with the problem on nearly 63 million acres subject to floods" and "to protect nearly 24 million acres . . . from critical erosion damage" as well as to irrigate 15 million acres and for other purposes. Obviously, the Small Watershed Program as envisaged by these supporters will be a large undertaking in the future. Here is another example of the relevance of our comments in Chapter 4 to the effect that many public research and investment programs in agriculture tend to increase productive capacity at the same time that other programs seek to restrain existing capacity from its full output.

All of this is to say that even a present attempt to measure the uncompleted soil conservation job involves working toward a poorly defined, and probably moving, target. One cannot simply say, when so many acres now suffering from certain disabilities have been treated, the soil conservation job will have been done. Long before that stage is reached, new horizons will have opened up, at least to the proponents of conservation. Nevertheless, it seems desirable to make an effort at estimating the unfinished job as it looks at present.

Conservation "Needs" of Cropland

The CNI estimated that 272 million acres of cropland in the forty-eight contiguous states needed conservation treatment of some kind in 1958 (see Table 22). This is approximately two-thirds of the total 1958 cropland area. This total acreage needing conservation is subdivided ac-

---

[2] *Basic Statistics on the National Inventory of Soil and Water Conservation Needs*, Statistical Bulletin 317, U.S. Department of Agriculture, August 1962, p. 5.

cording to dominant and secondary problems, and in considering this information, it is important to keep in mind that, when two or more problems were of approximately equal importance, erosion was considered the primary problem; excess water, the next most important; unfavorable soil conditions in the root zone next; and adverse climate last. In nearly 60 per cent of the area requiring treatment, erosion was listed as the dominant problem; for nearly two-thirds of this area (or for more than a third of the entire area requiring treatment), erosion was the only problem. If the areas where erosion was a secondary problem are added to those where it was primary, almost three-fourths of all the cropland requiring treatment had some erosion problem. Unfavorable soil, either as the primary or as the secondary problem, was listed as the next largest problem; and adverse climate was the least.

Because of the dedication to soil conservation of most of the men who made these surveys, they would be expected to be fully alive to erosion and other conservation hazards. And, as we have previously noted, there is at least the possibility that they unconsciously over-estimated the seriousness of the conservation problem; at the least, one would not expect them to underestimate it.

There is nothing in the CNI to indicate the capability class of land on which these problems exist. That is, we do not know what proportion of the land needing treatment with only an erosion problem is in land-capability Class II, where the problem is only moderate; what proportion is in Class III, where the problem is more severe; what proportion is in Class IV, which is questionable for continued crop production, especially in view of the fact that this land is not needed

TABLE 22. Conservation needs on cropland, by type of problem, 1958

(Millions of acres in 48 contiguous states)

Dominant problem in order of priority	Total needing treatment	No secondary problem	Secondary problem[a]			
			Erosion	Excess water	Unfavor- able soil	Adverse climate
Erosion (e)	161.4	99.2	—	9.0	28.4	24.8
Excess water (w)	59.9	33.1	10.1	—	16.1	.6
Unfavorable soil (s)	36.4	11.7	20.0	3.4	—	1.3
Adverse climate (c)	14.1	3.7	9.4	.1	.8	—
Total	271.8	147.7	39.5	12.5	45.3	26.7

[a] The priority of problems, ''when two kinds of limitations that can be modified or corrected are essentially equal,'' was given as they are listed here.

SOURCE: Data from *Basic Statistics of the National Inventory of Soil and Water Conservation Needs*, Statistical Bulletin 317, U.S. Department of Agriculture, August 1962, Table 28, pp. 141–45.

on a national basis to produce agricultural commodities; and what proportion is in Class V or worse—land which is unsuited for continuous crop production in any case. And similarly for lands with other problems, we lack information on the seriousness of the situation. Obviously, the class of land involved will make a great deal of difference as to how serious these problems are and as to the probable cost of treating them. Although the CNI presents statistics on the area of land requiring treatment by states and by counties, such data have not been mapped in any form so we can easily see where these lands are located. In these respects, as in others we have noted, the full information available from the CNI is not easily understood, nor did the basic surveys get all the information desirable for an understanding of the seriousness of the problems involved.

Neither the CNI nor any other report we have found provides details as to the kinds of conservation practices needed on these lands. In a policy report issued in May 1962 the Department of Agriculture says about such practices: "Measures included are those which experience shows are likely to be used in meeting the types of conservation problems inventoried. Some land treatment measures will need to be applied more than once during the next 20 years on some of the lands involved."[3] This report is not more specific as to the nature of these treatments, but it does go on to say: "About $33 billion would be required for conservation practices on cropland to solve problems caused by erosion, excess water, unfavorable soils, or adverse climate." After considering the extent to which public assistance programs might help farmers carry out these conservation practices and the current investment by farmers in conservation practices out of current farm incomes, the report concludes: "It appears that the attainable rate of public and private investment during the next 20 years would meet about one-third to one-half of the estimated cost of the conservation needs. Programs for conservation should continue to be reviewed with the objective of securing the greatest benefit from the investment of limited public and private funds."

The report implies, but does not state specifically, that these public and private investments are economically justified. Considerable doubt might exist on this point, however. An added $33 billion of investment in cropland—and other sums in pastureland and range and forest land, which we consider later—on top of the heavy capitalization, or over-

[3] *Land and Water Resources—a Policy Guide,* U.S. Department of Agriculture, May 1962 (slightly revised, September 1962), p. 61.

capitalization of agriculture may be dubious national policy. This report estimates future cropland needs for the nation on the assumption, among others, that future crop yields will be based on a continued application of conservation practices at about the past rate. However, since the nation has had burdensome agricultural surpluses, and in this report we conclude that it will continue to have them, the wisdom of a policy which adds to those surpluses is open to question. Possibly some different line of production adjustment and of soil conservation might be cheaper and also adequate.

One item in the national economic framework given to those who made the CNI was the estimate that the total cropland area would be 6 per cent greater in 1975 than in 1951–53. In spite of this basic estimate, the CNI reached the conclusion that net crop acreage would decline by about 11 million acres, or nearly 3 per cent. It was estimated that a larger acreage would come out of cropland, but that withdrawal would be largely offset by new land going into crops. By the time the basic statistics from the CNI were published in August 1962, the Department of Agriculture had reversed itself on the acreage of cropland needed, and stated that the need in 1980 was for 51 million acres, or 11 per cent less than in 1959.[4] While these figures are not exactly comparable with the CNI assumptions, since neither the base period nor future target date are exactly the same, they do represent a drastic change in outlook. Since we had estimated in our book, *Land for the Future*, published in 1960, that a reduction in crop acreage could be made, we naturally feel that the Department's later figures more accurately represent the situation.[5]

The Department's 1962 *Policy Guide* shows gross acreage shifts from one use to another, as well as net acreage shifts; but it does not include any data on classes of land that will be included in such shifts. If by some means the net shift out of cropland could be concentrated on the poorest land, then all of the land in capability Class V through Class VII and about half of that in Class IV which was used for crops in 1958 could be taken out of cultivation. One may logically assume that much of this land required conservation treatment in 1958, although precise information on this point is lacking. Certainly such land once needed major conservation attention, but we do not know how much of it may be included in the CNI's "adequately treated"

[4] *Ibid.*

[5] Clawson, Held, and C. H. Stoddard, *Land for the Future* (Baltimore: The Johns Hopkins Press for Resources for the Future, Inc., 1960).

category. It is impossible to concentrate all the acreage shifts out of cropland on the very poorest lands; some good land will be required for urban and other uses into which some cropland will go. However, substantial conservation gains could be achieved by careful selection of the lands to be retired because they are no longer needed. Many farms contain both good and poor land, but a farmer may ..esitate to reduce his already small crop acreage, even though some of his cropland is poor; and if his whole farm is retired, then the good land goes out of crops along with the poorer. Farm adjustment programs conceived and operated on an area or regional basis could almost surely increase the opportunities for taking poor land out of crops and concentrating crop production on the better land. This would have major importance to the soil conservation programs.

Conservation "Needs" of Privately Owned Pastures and Ranges

Conservation of pastures and range land involves consideration of the kind and density of plant cover, as well as of the soil itself. The two may be interrelated, since a thin plant cover is likely to mean an accelerated erosion hazard; but plant cover may be as dense as the climate will sustain, thus reducing erosion hazard to a practical minimum, and yet be an unproductive cover for livestock grazing. The old problem of maintaining basic productive capacity, or of increasing it, or of increasing annual inputs into an established input-output relationship again arises here. These considerations are especially acute for plant cover as contrasted with soil preservation. For one thing, changes in plant cover, especially improvements, are easier to make than are changes in soil conditions.

Of the total range and pasture area, roughly three-fourths was reported by the CNI as needing treatment of some kind. (See Table 23.) Almost half of this was reported as needing protection only; it can be assumed that, for the most part, this is a matter of preserving basic productive capacity in either soil or plant cover, or both. For the other half of the area needing treatment, the problem is either establishment or improvement of plant cover. While this may include some preservation of presently existing basic productive capacity, for the most part it probably includes additions to basic productive capacity.

Various protection measures are needed; some land requires more than one. The total of the enumerated protection needs approaches twice the total area reported as needing protection, indicating that, on the average, such lands require two practices of protection. The largest protection need, including almost 90 per cent of all land requiring protection, is against overgrazing. This is purely a management problem, primarily a matter of adjusting numbers of livestock and season of use to grazing capacity of the land; hence, it is very much a matter of annual inputs into a known or established input-output relationship. Investment in fences, water, and other facilities may simplify livestock management and make it possible to secure a larger amount of grazing without at the same time leading to overgrazing; but these too are more a matter of management than protection of basic productive capacity. The other protection needs are likely to be more for preservation of basic productive capacity of either soil or vegetation, but they may also involve some measure of increased input into an existing input-output relationship.

TABLE 23. Conservation needs on privately owned pastures and ranges, 1958
(Millions of acres in 48 contiguous states)

Land-use situation or conservation need	Acreage of total area	Per cent of total area
Not needing treatment	133	27
Needing treatment	364	73
Total pastures and range land	497	100
Needing treatment:		
Establishment of plant cover	72	14
Improvement of plant cover	107	22
Needing protection only	185	37
Subtotal	364	73
Needing protection only:	(185)	(37)
Need protection from overgrazing	163	
Need protection from fire	73	
Need protection from erosion	32	
Need protection from rodents	11	
Need protection from encroachment of plants	57	
Total enumerated protection needs	336	
Water management needed:		
Excess water	11	
Water conservation	23	
Total water management needs	34	

SOURCE: Data from *Basic Statistics of the National Inventory of Soil and Water Conservation Needs*, Statistical Bulletin 317, U.S. Department of Agriculture, August 1962, Table 30, pp. 149–50.

In addition to the protection and other needs, some water management problems were noted. For a rather small acreage, dispersed more or less over the entire nation, the water problem was one of excess water, or flooding. For about twice as large an acreage, concentrated primarily in the drier western states, the problem was one of conserving scarce water supplies.

As with other aspects of the CNI, no information is available as to the classes of land involved, nor as to the seriousness of the problem. We know, however, that roughly a third of all pastures and ranges is in capability Class I through Class IV. Is this the same land as the one-fourth of the pasture and range area which CNI reported does not need treatment? Or does some of this possible cropland need treatment, even for continued use as pasture and range? Somewhat less than a fourth of the range was capability Class VII, which by definition cannot profitably be improved by seeding, liming, fertilizing, water control, and the like. Where is this land, in the estimates of conservation needs of pasture and range? Is it listed as "not needing treatment" because it is uneconomic to treat? Or is it included in the area needing treatment where it has serious erosion in process? These are but some of the practical problems that arise when one considers the data in Table 23, but the CNI provides no basis for their answers. As with much of the other CNI data, no maps showing the geographic distribution of the various land classes and treatment groups are readily available.

The U.S. Department of Agriculture has estimated that the necessary conservation treatments on pastures and ranges will cost $10.5 billion.[6] The report implies, but does not state, that such treatments are economically justifiable. Considerable doubt may be raised on this point, however. In addition to the points made previously about cropland, to the effect that agriculture may already be seriously overcapitalized and that additional productive capacity is not now urgently needed, there is also the fact that much range land and pastureland does not respond well to various treatments, either in the physical or in the economic sense. Low rainfall, poor soils, and other factors often make the physical response small and slow; and the forage produced from ranges and pastures often has only a modest value per unit. It is true that in some cases erosion on range lands of low productivity may cause serious damage to downstream lands on the same watersheds, thus justifying measures which the grazing land alone would not support.

[6] *Land and Water Resources—A Policy Guide, op. cit.*

Conservation "Needs" of Private Forests and Woodlands

The conservation problem of forests and woodlands is similar to that of pastures and range land, in that there is a problem of maintaining the productive capacity of the vegetation as well as that of the soil. Even more than in the case of range, forest productivity is the result of actions taken in the relatively distant past. Even for pulpwood forestry operations in the South, in which there is relatively short rotation, the trees cut are from 20 to 40 or more years old; for many sawlog forestry operations, the trees cut are 100 or more years old. If a forest area is to yield a sawlog harvest of 80-year-old trees next year, or twenty years from now, this can only be possible if trees nearly 80 or 60 years old, respectively, are now standing and growing. Forest and woodland areas, no less than cropland and pasture-range areas, have the same problems of maintaining existing basic productive capacity, of adding new basic productive capacity, and of varying inputs into an existing input-output relationship; the estimated conservation needs must be considered in light of these distinctions.

The CNI information on the conservation needs of nonfederal forest and woodland areas was assembled primarily by county needs committees composed of local representatives of the Soil Conservation Service, of farmers, and of other local people, but with some help from state committees. The committees were instructed to use all possible sources of information and help, but the published results of their work show so many serious discrepancies—particularly from state to state—that one cannot help wondering if many of these committees were better informed about cropland and farming operations than about forests.

Of the total reported area of privately owned and state-owned forests and woodlands of 440 million acres, CNI reported that 200 million acres, or 45 per cent, did not need further treatment and 241 million acres, or 55 per cent, did need further treatment. (See Table 24.) These figures seem to reflect a healthier situation in the forests than on the range or croplands, since neither of the latter had so large an acreage which required no further treatment. In this connection, it should be noted that the Forest Service in 1958 included 358 million acres of privately owned commercial forest and 27 million acres of publicly, but not federally, owned commercial forest.[7] Thus, the CNI

---

[7] *Timber Resources for America's Future*, Forest Resource Report 14, Forest Service, U.S. Department of Agriculture, January 1958.

TABLE 24. Conservation needs of private forests and woodlands, 1958

(Millions of acres in 48 contiguous states)

Land-use situation or conservation need	Acreage of total area	Per cent of total area
Not needing treatment	200	45
Needing treatment	240	55
	—	—
Total forests and woodlands	440	100
Needing treatment only:		
Establishment and reinforcement of timber stand	69	16
Improvement of timber stand	160	36
Erosion control	12	3
	—	—
Subtotal	241	55
Need additional protection of timber stand from:		
Fire	252	
Insects and disease	207	
Animals, including rodents	82	
	—	
Total additional protection needs	541	

SOURCE: Data from *Basic Statistics of the National Inventory of Soil and Water Conservation Needs*, Statistical Bulletin 317, U.S. Department of Agriculture, August 1962, Table 31, p. 151.

must have included at least 55 million acres of noncommercial woodland or noncommercial forest. There is nothing to show in the discussions of areas requiring treatment in which groups these noncommercial lands were included.

Of the forests and woodlands judged to need further treatment, improvement of timber stand included the largest area, or 160 million acres, which is 36 per cent of all forests and woodlands, or 66 per cent of this land needing treatment. Establishment of timber stand was next most important in terms of area, with 69 million acres, or 16 per cent of all forest land, or 29 per cent of the area needing treatment. Although it is possible that these treatments were visualized primarily as a means of preserving existing basic productive capacity, it seems more probable that their major purpose would be to build additional basic productive capacity. In the case of forest products, in contrast to that for agricultural commodities, the outlook is for a greater demand than can readily be supplied from the existing productive capacity of forests, hence the rationale for public assistance in building such capacity is better. However, a generally greater future demand for forest products does not necessarily mean that all measures to increase output are economically justified. Although not explicitly stated in either the CNI instructions or published reports, it seems likely that the erosion problem of forests is not serious except for the relatively small area, 12 million acres, on which it is explicitly listed as a problem. For this small area, the problem probably is primarily one of maintaining the present basic productive capacity.

The reasonableness of the foregoing figures can be judged only when the discrepancies and gaps in details are examined. The Forest Service's review of timber resources had estimated that 51 million acres of forest land were "plantable" in 1952. The difference between this and the 69 million acres requiring establishment or reinforcement of timber stand, as shown by CNI, is considerable; state-by-state comparisons are not easily possible from the published data, but regional differences are large, in spite of some lack of comparability of regions. But these differences may not be of major concern; one's caution about the CNI results arises from the data on additional protection needs.

The CNI shows three classes of additional protection requirements for forest and woodland areas. The instructions for the CNI clearly stated that these were to include areas requiring additional protection, to bring the degree of protection to a high or fully adequate level. The area estimated to require additional fire protection is greater than the whole area requiring treatment—an apparent contradiction in terms. If the area requires additional fire protection, can it be considered as not needing treatment? This apparent discrepancy in national totals is far worse when data for states are examined. In Maine, for example, the entire forest area is reported as requiring better protection from fire, although 57 per cent of the forest area is reported as not needing treatment; a somewhat similar situation was reported for Kentucky (where actually more area was reported as needing better fire protection than was reported in forests and woodlands) and in Arkansas; and there were lesser, but still considerable, divergences in other states. In another contrast, Washington and California reported somewhat more area requiring better fire control than was reported as requiring any treatment; while Oregon reported no acreage at all requiring better fire control. These data certainly lead to the suspicion that radically different standards were applied by state and county needs committees. It is difficult, if not impossible, to make direct comparisons between the CNI and the Forest Service's review of timber resources as far as additional fire controls are concerned, since the definitions of adequacy of control do not seem directly comparable.

The CNI reported that 207 million acres—an area nearly as large as that requiring fire protection—required better protection from insects and disease. Again, curious relationships exist in state data. Vermont reported slightly more land requiring additional protection from insects and diseases than its whole forest and woodland area. Ohio and Arkansas reported all their areas requiring more protection, and Missouri and Florida reported almost all their total forest and woodland areas requiring additional protection—in every case, far more area than was reported as in need of treatment. In complete contrast, Pennsylvania, Maryland, Wisconsin, West Virginia, Tennessee, Oklahoma, Washington, Oregon, and California reported no acreage which required better protection against these hazards. It is hard to understand why Minnesota reported that most of its forest land required better protection, while Wisconsin reported that none needed better protection; or why Virginia, North Carolina, and Kentucky reported most of their acreage in need of better protection while West Virginia and Tennessee reported no acreage needing protection; similarly for Texas, which reported a major need, and Oklahoma, which reported none; and it strains one's credulity to think everything is as good as it could be in the forests of the three Pacific Coast states. Direct comparisons between the CNI and the Forest Service figures on insects and disease are impossible, since CNI reports acreages requiring better protection while the Forest Service reports annual growth losses due to insects and disease. However, one cannot avoid the conclusion that large discrepancies exist between these two sources of data.

As in other aspects of the CNI, no data are available as to the class of land on which the additional protection is needed, nor as to the seriousness of the problem. Nor are maps readily available showing where the land lies that was reported as requiring treatment. While these deficiencies are common to all the CNI data, the data about forest and woodland areas are even less reliable than that about either cropland or pastures and range land.

The U.S. Department of Agriculture estimated in 1962 that an additional $6 billion "would be necessary for establishment or improvement of farm woodland and commercial forests."[8] These estimates exclude the investment required on federally owned forests. Charles H. Stoddard had previously estimated that an investment of $2.1 billion was needed to improve all forests, including federally owned ones, in order to achieve the output goals postulated by the Forest Service.[9] These two estimates are not precisely comparable, partly because

---

[8] *Land and Water Resources—A Policy Guide, op. cit.*

[9] Stoddard, "An Estimate of Capital Needed in Forestry to Meet Projected Timber Requirements for the Year 2000," *Journal of Forestry*, Vol. 56, No. 7 (July 1958).

Stoddard includes a larger area, partly because the purpose of the investment is somewhat different. Whatever the precise figure, the case for further investment in forest productivity is better, as a general national policy, than is the case for more investment in cropland and in range land. This does not, of course, endorse every proposed investment in forests and woodlands as economically sound.

Remaining Conservation Problems on Federal Land

The growing demand for various uses of federal lands plus the incomplete status of past conservation efforts combine to provide a substantial backlog of conservation work. Much of the needed conservation work relates to forest and range vegetation, rather than to conservation of the soil as such. Moreover, it seems probable that further increases in demand for the use of these lands will require still greater conservation programs in the future, even if present targets are reached. Although the pricing of products from federal lands, current expenditures for management, and investment programs generally are not based upon careful economic analysis, it seems probable that the level of investment is gradually coming nearer to that which such economic analysis would indicate as optimum.[10] We may, therefore, regard present programs as only a step in the continued management of these public lands.

The Forest Service has presented a long-run development program for the national forests, together with a more specific program for the next ten years.[11] The 10-year program contemplates an expenditure for the decade of $2.5 billion, most of which would be capital investment, and a recurrent annual expenditure for protection and management of these resources of $185 million at the end of the period. While this program gives greater dollar emphasis to improving management capability than to conservation as such, and more to increasing basic productive capacity than to maintaining present capacity, its total objective is one of conserving and improving the productive capacity of the soil and associated plant cover. The general program is subdivided into parts dealing with water, timber, range, recreation, and wildlife resources, as follows:

[10] Marion Clawson and Burnell Held, *The Federal Lands: Their Use and Management* (Baltimore: The Johns Hopkins Press for Resources for the Future, Inc., 1957).
[11] *Development Program for the National Forests*, Miscellaneous Publication 896, Forest Service, U.S. Department of Agriculture, November 1961.

In addition to the general objectives for increasing the quantity and maintaining the quality of the water flowing off the national forests, the plan for the 10-year period includes 9,000 miles of gully and channel stabilization, 1.3 million acres of sheet erosion control, 10,000 acres of dune and blowout stabilization, erosion control on 13,000 miles of substandard roads and trails, 5,600 acres of water spreading, 410 structures for flood prevention, and 160 stream pollution control projects.

For timber resources, the long-range goal is to step up the timber cut to 21 billion board feet by the year 2000, compared with a cut of 11 billion in 1964. Various measures are proposed for the 10-year program that will permit an increase of the total cut while, at the same time, the productive capacity of the forests is maintained or increased. In addition, 4.4 million acres of nonstocked and poorly stocked plantable lands will be seeded or planted, and over 11 million acres will be substantially improved by plantation care, pruning, weeding, thinning, release cutting, reinforcement planting, and planting new burns.

Most of the 10-year program for range resources is for better management of livestock on the range, but 4 million acres are scheduled for revegetation or control of noxious and poisonous weeds.

The part of the program dealing with recreation resources contemplates a long-term use of 635 million visits by 2000, compared with 92 million in 1960 and with 123 million in 1963; and use is expected to double during the ten years of the program. Numerous specific measures are proposed to deal with this increased use and at the same time preserve the general attractiveness of the recreation areas. Many specific facilities would have to be installed.

In addition to taking an inventory of the wildlife habitat resources and developing better plans for them, specific measures for the 10-year program include improving food and cover on 1.5 million acres, developing openings and in other ways manipulating vegetation in favor of wildlife on 400,000 acres, and improving 7,000 miles of fishing streams and 56,000 acres of lakes in various ways.

Protection against insects and disease, fire, and other damage is a basic part of both the long-range program and the 10-year program, with numerous specific measures proposed. The building of additional roads and trails and improvement of existing ones would be a major part of the whole program, taking almost half of the total projected expenditure.

The Bureau of Land Management is currently developing a long-run soil conservation and resource improvement program.[12] Conservation needs for lands under the agency's control include brush control on 15 million acres, range reseeding on 12 million acres, and extensive fencing of the ranges and livestock water development. Some of these measures would maintain present basic productive capacity, while others would increase it, and some would facilitate management of the land.

These federal land programs are more concerned with management, or with current inputs and current outputs, than they are with preservation of basic productive capacity; and more concerned with the productive capacity of the plant cover than with that of the soil as such. Nevertheless, they do have important soil conservation results. Land covered with dense, healthy vegetation erodes much less easily than does land covered with sparse and unthrifty vegetation—whether the predominant vegetative cover is trees or grass. Several of the specific programs do deal with erosion problems as such, however; and the total effect of these various programs is to improve the soil conservation status of these lands.

Summary

In a review of the available statements about the remaining soil conservation job, two major impressions stand out:

1. The nation knows even less about where it is going than about where it has been on soil conservation generally. To some extent, this is natural; hindsight is always easier than foresight, but the discrepancy here seems needlessly large. The statements of general objectives, of specific goals, and of costs and benefits for future soil conservation are much less precise than they could be.

2. The public agencies concerned with soil conservation seem to contemplate more of the same—a continuation, with only modest changes, of past programs. This seems especially marked for those programs dealing with soil conservation on privately owned land. There has been no critical re-examination of the objectives, methods, and timing of programs, as far as we can observe. The potentials of adjustments in land use inherent in the ongoing agricultural revolution do not seem to enter into soil conservation planning.

[12] Correspondence from Eugene V. Zumwalt of the Bureau of Management, March 6, 1964.

In view of the rather substantial public costs involved in the federal soil conservation programs, it would seem that the interested public has a right to insist upon a more careful re-examination and re-formulation of these programs than has been undertaken to date. Until there is better data on what has been done and what the current situation is, and a clearer idea of what comes next, estimates of numbers of acres or sums of dollars have relatively little meaning.

Chapter 10

FORCES IMPEDING AND
PROMOTING SOIL CONSERVATION

W<small>HY</small> should any farmer, landowner, or society as a whole tolerate any erosion? Perhaps no one would attempt to defend the extreme position implied in this question for some erosion is a normal geologic process. But why tolerate any accelerated erosion? Some might answer that a measure of accelerated erosion is unavoidable if any productive use is to be made of soil and other resources. There may well be some conservationists who would retort that no use or profit can justify the destruction of basic soil and other resources. The economist would surely say that erosion must be considered in relation to the costs of controlling it and the returns which an erosion-causing land use makes possible. This view would almost surely be rejected by the truly devoted conservationist. Differing objectives and goals certainly would affect the attitude toward specific means.

In the foregoing chapters, we have traced, in a general way, the history of soil conservation as a major social movement in the United States; and we have tried to take an inventory on where we are now, both as to job accomplished and job yet to be done. Before we undertake, in the third major part of this book, to look ahead for a few decades, we shall try in this chapter to review the major forces impeding and promoting soil conservation.[1] To some extent, this will require going over ground that we have covered previously, but it will be from a different viewpoint and thus may be constructive enough to warrant repetition.

In line with the framework set forth in Chapter 1, there are at least three dimensions of a realizable goal for soil conservation. These are:

---

[1] So many reports deal peripherally with the topic of this chapter that we will not attempt to cite them all. However, we will cite at least some of those which are concerned more directly with the subject.

253

1. To maintain the productivity of the basic soil resource, by holding soil erosion to the level where the formation of new soils from basic material at least balances soil loss.

2. To increase basic soil productivity, wherever the social or total gain exceeds the total cost to the farmer and to society.

3. To encourage annual inputs up to the optimum or maximum profit level.

These different objectives or goals are theoretically distinct but, in practice, many specific programs tend to have two or more effects. Without at this point trying to say which of these is the more important, or which may even be undesirable, we shall simply consider the forces that will impede or promote one or more of these general goals.

Farmers' Awareness and Response

One basic factor in the achievement of soil conservation is the attitude of the farmer or other landowner or occupier. Several more or less distinct issues or steps exist in the continuum from no action to complete action.

First of all, the farmer or other landowner may not recognize that soil erosion is taking place. When gullies have become large, they are so obvious that no one who works the land can be oblivious to them; and really spectacular wind erosion such as part of the country experienced in the thirties is equally hard to ignore. But a great deal of more moderate wind or sheet erosion, which in time could do very severe damage to the land, simply may not be seen by the man who lives on and works the land. As Hugh H. Bennett and others said repeatedly, there was an amazing lack of knowledge about even the existence of soil erosion in the early thirties. In part, this was because it is hard to measure actual soil loss accurately under usual farm conditions. By now, various formulas have been devised which rather accurately estimate what the soil loss would be, if it could be measured exactly. Today, inability to recognize erosion, or denial that it exists, is far less common than it once was.

Next there is the man who recognizes that some erosion is taking place, but who does not regard it as a serious matter. Farmers and others, conceding that some soil loss is taking place, may feel that it is normal or unimportant; may remark that "a little soil is washing but so what?" They do not distinguish between relatively low rates of soil loss, such as those which often occur even on undisturbed soils, and accelerated soil losses which in a relatively few years will lead to serious impairment of soil productivity. This attitude is often compounded by an ignorance about soil profiles, soil formation, and the difference between topsoil and subsoil. The rancher who viewed serious earth slides resulting from continued burning of native vegetation on steep slopes with the remark, "Don't worry, it is like that all the way through to China," may have been unusual; but indifference to soil losses was, and to some extent still is, common.

A further step is taken when the farmer or other land user realizes that something is wrong—although he may not be fully aware as to just how serious the danger is—but does not know just what to do about it. He may be unaware of the role that terraces, or grassed waterways, or cultivation on the contour, or any of various other soil conservation practices might play in reducing soil loss on his farm. Or he may be aware of the practices but not know how to apply them. Soil conservation practices which now seem simple, such as plowing and cultivation on the contour, did not seem so simple before they were in general use. Farmers gain information from many sources, but the successful use of some practice by a neighbor is often one of the most effective. Even in a country like the United States, it is not a simple matter to start a wholly new practice. Certainly, soil conservation measures of various kinds are no exception.

At a still later stage, when the farmer or landowner is told about one or more soil conservation measures by someone—whether by a specialist from a public agency or by neighbor—he rejects the practice for one of several reasons. He may think it unnecessary, that something else less drastic will do the job; or he may feel that it will be ineffective; or it may seem impractical in view of other necessary farm measures or practices; or there may be other reasons. Proposals for farming on the contour—"around the hill"—often aroused much skepticism, not to say derision, in many areas when they were first proposed. The man who had spent all his life learning and practicing to plow an absolutely straight furrow was shocked to be urged to plow a deliberately curving one. There was much concern over short rows which were necessary when terraces on the contour could not be parallel because of varying slopes. Many other illustrations could be cited. One must concede, too, that in some situations, especially at the early stages of the soil conservation programs, the specialists often did not know what to advise or how to apply properly the general measures.

All of these general comments about soil conservation apply equally well to a consideration of the conservation of forest and range vegetation. Often little was known about the requirements of the forest for successful regeneration after harvest, or the limitations on grazing needed to preserve the productivity of the range plants. Frequently, deterioration of range was unrecognized or denied, partly because it was gradual. Measures to preserve or rebuild productivity about which little was known often were strongly resisted.

In all land use situations, there is a considerable inertia or resistance to change. True, in the United States, this has been very much less than in many countries where the son was taught to follow closely the practices of his father, who in turn had learned them from his father. But U.S. farmers have traditionally been conservative and slow to change; for a variety of reasons, many have been slow to adopt improved practices, even when they would concede their theoretical superiority.

The traditional roles of education and research of the public agencies are well-suited to deal with all these problems. Where the problems are primarily lack of knowledge or lack of willingness to use the best available knowledge, the customary approaches eventually may yield results. For soil conservation, however, it must be recognized that the problems are as much social or human as they are physical or technological. It is not merely a matter of knowing how erosion should be controlled in an area, or of telling the farmer about it; it is also a matter of convincing him, which often requires a shift in approach from that of the specialist. The point of view of the farmer and his criteria in reaching a decision may be quite different from those of specialists in deciding upon their recommendation.

These general points may be briefly illustrated in some specific studies relating to agricultural innovators in general. Eugene E. Wilkening found that the "reasons for not approving the adoption of specific improved practices [of a general agricultural production type, not specific to soil conservation] were of four general types: (1) failure to recognize the advantages or the effectiveness of the improved practice; (2) lack of means for implementing the practices including land, labor, or capital or rental arrangements; (3) dissatisfaction with particular aspects of the practice including inconveniences and changes in operations; and (4) conflicts with other operations or activities."[2] In a somewhat similar study in Wisconsin, the same author found that the process of change was a gradual one, with farmers first hearing

[2] Wilkening, *Acceptance of Improved Farm Practices in Three Coastal Plain Counties,* Technical Bulletin 98, North Carolina Agricultural Experiment Station, Raleigh, N.C., May 1952.

about a new practice, then accepting it as a "good idea," then accepting the practice on a trial basis, and finally adopting the practice completely.[3] Ward W. Bauder later found similar relationships for the acceptance of new farm practices in the cash-corn region of Illinois.[4]

In considering the rate of adoption of improved farm organization plans that could be shown to increase farm incomes significantly in the Piedmont area of South Carolina, Calvin C. Taylor and Thomas A. Burch found five major obstacles: (1) age and physical handicaps, with older farmers unable or unwilling to make changes; (2) limited education and training, which is often reflected in pessimism and conservatism toward technological advance; (3) consequences of a rapidly changing agricultural economy, which has often left the older farmers more or less stranded; (4) inadequate resources to carry out the profitable changes; and (5) poor coordination of farm and off-farm employment.[5] James H. Capp found in Kansas that the youngest and the oldest farmers were least responsive to adoption of improved farm practices, that those qualities which made for high membership participation in church and farm organizations also led to relatively rapid adoption of new practices, and that the same was true for strong interest in community affairs generally.[6]

Some studies of this general type have led George M. Beal and Joe M. Bohlen and other students of the problem to establish several more or less distinct stages in the process of agricultural change: awareness, where the farmer has heard about the practice or method but has little knowledge; interest, where he desires more information; evaluation, where he makes a mental trial of its advantages and disadvantages to him; trial, when he actually tries out the new idea and when he needs specific information; and adoption, or continued use on a large scale.[7] These same authors, and others, have classified farmers

[3] Wilkening, *Adoption of Improved Farm Practices as Related to Family Factors,* Research Bulletin 183, Agricultural Experiment Station, University of Wisconsin, Madison, Wis., December 1953.

[4] Bauder, *Influences on Acceptance of Fertilizer Practices in Piatt County, Illinois,* Bulletin 679, Agricultural Experiment Station, University of Illinois, Urbana, Ill., December 1961.

[5] Taylor and Burch, *Personal and Environmental Obstacles to Production Adjustments on South Carolina Piedmont Area Farms,* Bulletin 466, South Carolina Agricultural Experiment Station, Clemson, S.C., December 1958.

[6] Capp, *Personal and Social Factors Associated with the Adoption of Recommended Farm Practices among Cattlemen,* Technical Bulletin 83, Kansas Agricultural Experiment Station, Manhattan, Kan., 1956.

[7] Beal and Bohlen, *The Diffusion Process,* Special Report 18, Agricultural Extension Service, Iowa State College, Ames, Iowa, March 1957.

into broad groups; innovators, who can afford to take risks and are willing to do so; early adopters, generally younger, better educated, and more alert than the following groups; early majority, slightly above average in various personal factors which lead to adoption of new practices; majority, or the larger portion of the total farm population, which comes along after the way has been blazed by the foregoing; and non-adopters, who lag the farthest and who are the oldest, least educated, and who participate least in all kinds of community organization.

These general ideas find specific expression in the rate at which farmers adopted hybrid seed corn in different states (Figure 41). Hybrid seed corn represents a practice that is unusually easy and profitable to adopt. A farmer who is growing corn continues to prepare, plant, and cultivate the land, and to harvest the crop as he has always done, and to use the corn as he always has; no major change in farm organization or operation is required. He has only to buy seed, which is often more expensive than purchase of single line seed or than using his own seed; but his harvest is greater, often by as much as 25 per cent from this one single practice. In many areas, it is true, use of the practice was impractical until locally adapted varieties were developed by seed breeders; but, once these were available to some farmers, the further spread of the practice was dependent almost entirely on the personal characteristics of the farmers. The general similarity of the adoption curve or pattern for the United States and for a few sample states is evident. In each case, the practice is begun by a few well-informed or venturesome farmers (the innovators); it is soon picked up by the majority of farmers; and a point is reached after which only a few laggards have not adopted the practice.

A number of reports contain information bearing upon this same general process as it relates to soil conservation. Claude C. Haren found that some land was abandoned in the southern Piedmont because of the age or poor health of the owner, or from inability to hire help or to rent land to neighbors, or because the farms or fields were too small, poorly located, or unadapted to modern machinery. In this locality, in general, abandonment of cropland usually means a large measure of soil conservation, since weeds, brush, and trees come in to provide protective cover.[8] H. O. Anderson and associates found in southwestern Wisconsin that various personal characteristics of the

---

[8] Haren, *Conservation Farming in Selected Areas of the Southern Piedmont,* ARS 43–120, Agricultural Research Service, U.S. Department of Agriculture, August 1960.

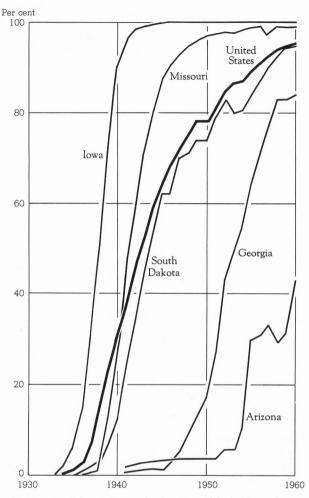

Figure 41. The pattern of adoption of hybrid seed corn in the United States and five selected states, 1930–60.

farmer often greatly influenced his willingness to adopt soil conservation practices, and that farm plans had to take into account the willingness and ability of the farmer to carry out specific measures.[9] They also found that time was a major factor—as the farmer had longer experience with a specific conservation measure, he not only used it more

[9] Anderson, P. E. McNall, and Buis T. Inman, *Progress in Application of Soil-Conserving Practices, Southwestern Wisconsin,* ARS 43–44, Agricultural Research Service, U.S. Department of Agriculture, April 1957.

but saw its advantages more clearly and was less disconcerted by its disadvantages.

In a study which drew upon the accumulated experience and research of its members, the North Central Farm Management and Land Tenure Research Committees of the North Central Region listed nine major obstacles to greater use of soil conservation measures. These were:

1. Reluctance of farm operators to change old methods of farming, insufficient skill to lay out the work, lack of accurate information on costs and benefits, and frequent changes in recommended conservation practices.

2. Organization problems on small farms where intensive crops are used to keep the family labor force gainfully employed.

3. Land holding and rental procedures that restrict the interest of either owners or renters to periods which are shorter than the time required to carry out a conservation plan, or to receive the benefits as great or greater than the cost of the improvements.

4. Reluctance of farmers to pay the out-of-pocket costs of making necessary changes in farm organization.

5. The time lag between cash outlays for conservation work and returns.

6. The desire of owners and operators for high current income.

7. Uncertainty as to future prices and weather conditions, and the possibility of losing a considerable part of the money invested in the animals that are needed to process hay and pasture into readily salable products.

8. Differences between maximum long-run income to the individual farmer and a socially desirable level of conservation.

9. Relationship of conservation to the general problem of better farming.[10]

Since the foregoing presumably were to include all major obstacles to conservation on farms in this region, the heavy emphasis upon personal characteristics and attitudes of farmers is particularly noteworthy.

In a series of studies, researchers at Iowa State College (now Iowa State University) have explored the soil conservation problem in that state, especially along the western edge which, as we noted earlier, every survey of soil erosion or conservation has shown as a major

---

[10] *Obstacles to Conservation on Midwestern Farms,* Bulletin 574, Agricultural Experiment Station, University of Missouri, Columbia, Mo., June 1952, p. 9.

problem area.[11] In spite of a conclusion by Ross V. Baumann and his associates that "several alternative ways are suggested by which old gullies can be controlled, new gullies prevented and the productivity of the soil maintained or improved" and that "the change from present to alternative systems of farming would be profitable," the fact is that progress in soil conservation in serious problem situations, especially along the western edge of the state, has been very slow. Various personal reasons of the farmers or institutional factors are largely responsible for the slow progress. Farmers simply do not think present erosion is as serious as do soil conservation specialists. While the farmers will concede that soil losses should be reduced, their objectives are far below those which the specialists think are the minimum acceptable ones; actual soil conservation achievements, in turn, are below their goals. One major factor is tenure arrangements (to which we will return later in this chapter); another is lack of capital to engage in livestock farming, or distrust about future income and security of investments in livestock enterprises. But, at least as important, and possibly more so, is the reluctance of the farmers to give up a system of cash corn farming for one involving major reliance on livestock. Also a factor is high taxes on land. The general reluctance of farmers to change finds expression in a number of specific objections to particular practices or methods, such as terraces or contour cultivation. If recommended soil conservation practices were to be adopted in this Iowa locality, its total agricultural output would decline somewhat. Here is an example of soil conservation being output-decreasing in effect.

Although the attitudes of farmers toward soil conservation—as toward change in general—have been a major factor impeding rapid adoption of new soil conservation measures, the effect of these attitudes gradually diminishes with longer exposure to the new measures

[11] In historical sequence, some reports in the series—all from Iowa State University of Science and Technology, Ames, Iowa—are the following:

John C. Frey, *Some Obstacles to Soil Erosion Control in Western Iowa*, Research Bulletin 391, October 1952.

Ross V. Baumann, Earl O. Heady, and Andrew R. Aandahl, *Costs and Returns for Soil-Conserving Systems of Farming on Ida-Monona Soils in Iowa*, Research Bulletin 429, June 1955.

R. Burnell Held and John F. Timmons, *Soil Erosion Control in Process in Western Iowa*, Research Bulletin 460, August 1958.

Lloyd K. Fischer and John F. Timmons, *Progress and Problems in the Iowa Soil Conservation Districts Program*, Research Bulletin 466, April 1959.

R. Burnell Held, Melvin G. Blase, and John F. Timmons, *Soil Erosion and Some Means for Its Control*, Special Report 29, August 1962.

and systems. Since there has been a major federal soil conservation program for thirty years, it would appear that continued reluctance to adopt recommended practices and systems might rest upon more than farmer attitudes. Thus, we now consider the relative profitability of soil conservation, and later will consider some of the institutional obstacles. Basic to our approach is the assumption that difficulties will be overcome, often quickly, if the farmer or other person concerned is very likely to receive substantial financial gains by doing so; or, conversely, that difficulties are likely to loom large so long as there is little prospect of financial gain from overcoming them. Difficulties are thus relative; even the most firmly entrenched attitude yields to the prospect of large financial gain, while trivial matters present major obstacles when the gains are uncertain or probably small.

Costs, Benefits, and Profits to the Individual Farmer

The farmer who undertakes soil conservation experiences different kinds of costs in the process. Some added costs are in the form of cash for annual inputs—more fertilizer, different seeds, and so on. Other added costs are for investment in such soil conservation works as terraces; the annual cost is a reasonable interest return on and amortization of the investment, plus some upkeep. Still other costs are in the form of income that is foregone or postponed; in the humid part of the United States, many farmers could rather easily solve their soil conservation problems by allowing the land to grow up to weeds, brush, and trees which would protect the land against erosion but at the same time reduce its economic output to zero or nearly so. A less extreme case is the conversion of cropland into pasture; the income from the sale of crops would be abandoned or reduced, but the production of livestock would be increased. Beyond these cash costs, many human or attitudinal costs are involved in soil conservation. The farmer who must violate all he has learned about plowing a straight furrow in order to farm on the contour may experience a severe psychological cost. Or the addition of livestock, with all that this means in terms of daily care and reduced freedom of time, also may involve a serious personal cost that is not easily measured in dollars.

Just as they have costs, farmers also obtain benefits of various kinds if they undertake soil conservation programs. The most basic benefit is the prevention of further deterioration of the soil, and the ultimate economic loss that this would mean. A soil conservation program will have an effect upon immediate cash income; the income may be lowered, temporarily or indefinitely, or it may be raised by the use of more inputs and a higher level of management. The primary income consideration is cash, either now or over the future, with future cash suitably discounted back to a present value. But some farmers, at least, get psychic income from well-kept fields, productive pastures and ranges, and well-stocked forests. These may not weigh as heavily as added income for most farmers, yet their importance should not be overlooked.

The timing of costs and returns from soil conservation programs is especially important. As we noted in Chapter 1, several kinds of situations are possible; but rarely will expenditures and benefits occur in exactly concurrent fashion. A typical situation, but by no means the only one, is an immediate expenditure, largely for investment purposes, against which expected future income must be weighed. Not only is the timing different for the expenditure and the income, but so is the certainty of each; the expenditure is direct and definite, the future income, unavoidably, is more speculative. Professional researchers often differ as to the most appropriate way of making comparisons—which interest rate to use, and the like. But far less is known about how farmers actually compare present and future expenses and income. There is good reason to suspect that farmers discount a relatively distant and perhaps somewhat uncertain future income much more severely than the researcher. Under any circumstances, an income stream beyond fifty years from the present has a very low present value, and one beyond twenty years from the present also has a limited present value. But perhaps farmers give no credit at all to incomes realizable fifteen or more years in the future, and give less to an income ten years or more in the future than a reasonable discounting process would suggest.

Another factor of great practical importance is the fixity of most input factors in many kinds of farming. Economic analysis works best for incremental adjustments of input—adding a little more labor, or land, or capital, to fixed or determined quantities of other factors. In the short run, many farmers can vary inputs only a little, if at all. They want a job and the only place they may be employable is in agriculture and right on the present farm; their personal labor supply may have little or no value except in farming.[12] Their "salvage value" for other occupations is low. On the other hand, they may be in a poor

---

[12] Glenn L. Johnson, "Supply Function—Some Facts and Notions," in *Agricultural Adjustment Problems in a Growing Economy*, edited by Earl O. Heady *et al.* (Ames: Iowa State College Press, 1958); and Glenn L. Johnson, "The State of Agricultural Supply Analysis," *Journal of Farm Economics*, Vol. 42, No. 2 (May 1960).

position to hire additional labor. They have a given area of land, often too small for its present purposes, and certainly too small to consider diverting some crop acreage to pasture or woods; yet renting or buying additional land, especially in incremental quantities, may be very difficult or impossible. Their supply of annual input capital is severely limited, as is their credit. In the longer run, many more possibilities open up—the farmer may die or quit, thus withdrawing his labor supply; he may sell his farm, thus opening up the possibility of land adjustment to someone else; he may buy some other farm when it is offered for sale; and his capital supply may either increase or decrease. But individual farms are rather inflexible, even in the long run. Agriculture as an industry perhaps has greater flexibility due in part to the fact that the number of farms can be decreased sharply if necessary. But the farm operator may be unwilling to make any significant change toward greater soil conservation during his lifetime or at least during his occupancy of the farm.

In order to take full economic advantage of a program of soil conservation on a farm, it is often necessary also to make changes in the other inputs. This may be especially true of managerial skill— frequently the most limiting input of all. A number of studies have shown how farms could be reorganized and operated differently, both to increase soil conservation and to increase current income. But this is almost always possible by the use of more inputs, or by a more intensive system of farm operation, which requires substantially more managerial competence. Under these circumstances, one may well question whether the higher income results primarily from the soil conservation program or primarily from the different system of farm management—one aspect of the more profitable system being more attention to conservation.

Closely related to all the foregoing is the length of the planning period which actually guides farmer decisions. Will any farmer make plans which extend beyond his reasonable life expectancy, unless he has a son ready to take over management of the farm? Given the older age of many farmers, and the fact that a ready replacement for the low-income farm is often not available, the planning horizon for many farms is very short. In practice, do many farmers operate with a planning horizon considerably shorter than their own probable life? Does soil conservation farming have to be concerned primarily with the next five years on the farm, less so with the five ensuing years, and hardly at all for periods beyond ten years? Possibly not for many profitable farms operated by moderately young and quite able farmers; but it may be unreasonably optimistic to count on longer periods for a larger number of small, poor farms, operated by older men.

A review of the available literature, as well as an examination of the actual developments in soil conservation, leads to a general conclusion: In the absence of public subsidy, much proposed soil conservation has a low profit potential, some is even negative, and only comparatively little promises to be highly profitable. To put it differently: With some exceptions, private soil conservation efforts generally do not promise large profits; rather, net gains will be modest in many cases, and substandard in others. If this general conclusion is true, then it goes far toward explaining the modest progress of past soil conservation, both in many regions and nationally, and it has great significance for the future.

One other general point is important: The land market, as nearly as we can tell, places relatively little value on soil conservation measures, or on a farm's conservation health or lack of it. That is, a well-kept farm, with a practical minimum of soil erosion, is likely to sell for somewhat more than a rather run-down one in the same locality with similar basic soil conditions; but the difference in price is not likely to be anywhere near as much as it would cost to rejuvenate the poorer land. Disheartening as this relationship may be to devoted soil conservationists, there is some rationale for it. Soil conservation depends heavily on the kind and competence of farm management; a well-kept farm might be sold to someone who would let it deteriorate, while a farm in bad condition might be bought by someone who would restore it. But, if it is true that soil conservation expenditures do not increase the selling price of the farm by as much as they cost, then this surely puts a serious dampening influence on soil conservation expenditures, especially those which will not pay off soon. This relationship tends further to shorten the already short planning horizon of many farmers.

The foregoing general statements are borne out by several specific studies. Arthur J. Coutu and his associates put the matter very well when they say: "There can be no single answer to the question: Does conservation pay?"[13] They go on to say: "Whether any specific practice will pay depends upon: (1) the length of the planning period; (2) dif-

---

[13] Coutu, W. W. McPherson, and Lee R. Martin, *Methods for an Economic Evaluation of Soil Conservation Practices,* Technical Bulletin 137, North Carolina Agricultural Experiment Station, Raleigh, N.C., January 1959.

ferences between yields, over the appropriate time period, when the particular practice is employed compared with not using the practice; (3) product and factor prices over the appropriate time period; and (4) discount rates." They explored the economics of a complete terracing program with contour cultivation and recommended runoffs and waterways, one with terracing and meadow outlets without contour cultivation, and complete terracing with contour cultivation and strip cropping, contrasting each program with what the situation would be without treatment. A major problem in economic evaluation is to obtain reliable estimates of probable yields with and without various conservation treatments. In view of this, the analysis was carried forward in terms of the best available estimates of yield differences, and also by calculating what yield differences would be necessary in order to make a specific program economically feasible. Substantial investments and somewhat higher annual costs were involved for each of the conservation programs, compared with none at all; and the general farm operations were more intensive with conservation programs than without them. The time of break-even, when the costs, including interest, had been recovered, were also calculated; and this varied from as few as three years in some circumstances to well over fifteen; but it was considered that fifteen years was about the practical limit of the planning horizon of farmers. While this study shows that recommended soil conservation practices are profitable under certain assumptions as to future conditions, they are not invariably so.

Somewhat similar results were obtained by S. W. Atkins in a study of West Tennessee:

> In the early years, however, net incomes under high conservation would be reduced below incomes obtained under low conservation. Net labor returns under high conservation on the case-study dairy-hog-cotton farm, for example, would be 10% less than net returns under low conservation in the bench mark period.
>
> If all costs (cash and noncash) are charged and future incomes are discounted at 6% per year, high conservation would not be profitable during the lifetime of present farmers under the assumptions of constant technology and constant prices. The sum of all present values of future incomes from high conservation would be 10% below similar incomes from low conservation for the first 50 years. For moderate conservation, the comparative income would be 3%

below that for the low conservation level of management. Omission of noncash costs, however, would improve the relative income position of the higher conservation systems.[14]

W. C. McArthur and John R. Carreker conclude for the Piedmont area of Georgia:

The selection of a [profitable] conservation plan for a given farm depends on several factors which include: (1) the physical features of the farm, including soil, slope, and degree of erosion, that affect the conservation practices required; (2) the amount of capital available to the farmer, and alternative investment opportunities; (3) the amount of labor available, and seasonal variations in labor requirements for different crop and livestock enterprises; (4) yields of different crops; (5) production costs per unit for different crop and livestock enterprises; (6) the different kinds of livestock that can be adapted to a given cropping system; (7) prices of different farm products; and (8) managerial skill and "know-how" of the farm operator and his ability to withstand wide variations in farm income. The most profitable combination of enterprises in a conservation farming system depends on: (1) the production response of individual crops to conservation practices, and (2) the prices of farm products compared with the costs or prices of factors used in production.[15]

R. N. S. Harris and his associates, in their general study of cropland reversion and soil conservation in the whole South, conclude:

A finding is that conservation expenses of eroding abandoned land in the Southern Piedmont are not covered by future benefits at interest rates likely to be appropriate for the individual farmer. Benefits exceed costs only if a low social discount rate of 2.5 or possibly 5 percent is appropriate. A fairly good case can be made for conservation for forest production at these low discount rates. No ready measures of reliability are available of physical and economic estimates underlying the analysis. However, on a judgment basis, it may be stated that the results establish fairly definitely that,

[14] Atkins, *Economic Appraisal of Conservation Farming in the Grenada-Loring-Memphis Soil Area of West Tennessee,* Bulletin 369, Agricultural Experiment Station, University of Tennessee, Knoxville, Tenn., October 1963, pp. 2–3.

[15] McArthur and Carreker, *Economic Analysis of Conservation Farming on a Cotton-Dairy Farm in the Piedmont Area of Georgia,* Bulletin N.S. 51, Georgia Agricultural Experiment Stations, Athens, Ga., March 1958, p. 40.

268 Performance and Evaluation

if conservation is to be undertaken, public assistance will be required since the measures pay only at low discount rates.[16]

Herbert Warren Grubb has made a detailed analysis of soil savings and net benefits in a number of specific situations in four southern states. He concludes:

Estimated net benefits were positive for nine of the twelve counties, but farm-to-farm sampling variability was so great that net benefits for only two of the twelve counties could be considered significantly different from zero. These findings hold for all interest rates (0.001 to 0.09) and time horizons (5 to 100 years) that were considered in this study. A larger sample of farms might have resulted in benefits being found statistically different from zero in more counties. . . . The results of this study are not strongly in favor of present soil conservation activities in the sense that it cannot be said with confidence that there exists positive net benefits for a large proportion of counties. . . . Benefits to terracing were *negative* on 17 to 23 of the [24 sample] farms (depending on interest rate and time horizon). Benefits to crop rotation changes were *positive* on 17 to 21 farms.[17]

Walter W. Pawson and his associates in their study of the Palouse area of eastern Washington concluded: "The findings indicate that farmers can adopt cropping systems that will maintain the productivity of the soil better than the prevailing wheat-pea system and that will produce as high or higher net income under the price and cost conditions considered."[18] The best cropping system depends upon a number of factors, including federal agricultural programs and their restrictions on wheat acreage. While the authors' conclusion follows from their data, the high-income cropping systems generally involve a more intensive level of operations and perhaps a higher degree of managerial skill

[16] Harris, G. S. Tolley, and A. J. Coutu, *Cropland Reversion in the South,* A. E. Information Series 100, Department of Agricultural Economics, North Carolina State College of the University of North Carolina at Raleigh, May 1963, p. 69.

[17] Grubb, *Individual and Aggregate Benefits and Costs of Soil Conservation in the South,* unpublished doctoral dissertation, Department of Agricultural Economics, North Carolina State College of the University of North Carolina at Raleigh, 1964, pp. 80–81.

[18] Pawson, *et al., Economics of Cropping Systems and Soil Conservation in the Palouse,* Bulletin 2, published co-operatively by the Agricultural Experiment Stations of Idaho, Oregon, and Washington (Moscow, Idaho; Corvallis, Ore.; and Pullman, Wash.), and the Agricultural Research Service, U.S. Department of Agriculture (Washington, D.C.), August 1961.

than low-income systems. The cropping systems on both types of farms are often similar. Moreover, it appears that under some assumptions, a system involving a higher soil loss per acre is more profitable than one involving a low loss (see page 67, Figure 58 of their publication). They do show that use of commercial nitrogen fertilizer can help maintain soil productivity and also be profitable under many circumstances.

In a study of more than a hundred farms located on slowly permeable soils of northeastern Illinois, E. L. Sauer and his associates concluded that conservation measures not only were effective in maintaining soils for future use but that they could contribute to net farm income.[19] However, this result was obtained by operating at a more intensive level of inputs and outputs, and perhaps requiring a higher level of managerial skill, for the conservation farming. More livestock were also needed to take full advantage of the greater pasture and feed supply resulting from the conservation farming; and, in many cases, lease adjustments were also necessary. The authors also point out that time is required for conservation investments to pay off. A limited basis for credit for conservation measures exists. This study is a clear example of the possibilities in some areas of more intensive farming—which can also mean more conservation in farming.

In a later study of this same area, two other authors, C. E. Harshbarger and E. R. Swanson, come to a different conclusion: "If the relationship between soil loss and yield is studied in isolation from changes in technique of production, a farmer on Swygert soils would sacrifice income by keeping soil losses at or below the acceptable level."[20] They followed the usual techniques of estimating annual soil losses by appropriate formulas, under different cropping systems, and of estimating farm income under each. However, they assumed that decreased yields caused by increased erosion would not be offset by increased use of fertilizer. Presumably, by using more fertilizer and perhaps by other practices, yields could have been maintained or possibly increased.

But they judged that such practices would have confused the effect of soil conservation with the possibilities of more intensive manage-

[19] Sauer, J. L. McGurk, and L. J. Norton, *Costs and Benefits from Soil Conservation in Northeastern Illinois,* Bulletin 540, Agricultural Experiment Station, University of Illinois, Urbana, Ill., in cooperation with Soil Conservation Service of the U.S. Department of Agriculture, June 1950.

[20] Harshbarger and Swanson, "Soil Loss Tolerance and the Economics of Soil Conservation on Swygert Soils," *Illinois Agricultural Economics,* Vol. 4, No. 2 (July 1964).

ment, hence they excluded it. For each soil situation, each discount rate, and each time period for planning, a cropping program of continuous corn gave the highest returns on this soil; but this program also resulted in the highest annual soil loss per acre.

In southwestern Wisconsin, H. O. Anderson and associates found that net incomes were higher on farms with a high degree of soil conservation than on farms with a low degree, and that they were higher the longer the farm had participated in a soil conservation program.[21] The differences were not great—about 7 per cent between groups of farms with high and with low degrees of conservation, but about 36 per cent between farms in the program less than three years and those in it ten or more years. No measure is given of income variation among farms, which might have been large; hence, no conclusions can be drawn as to the reliability of these relatively small differences in income. As in several other areas, the higher conservation is achieved by a more intensive level of farm operations; both expenditures for inputs and gross output per acre are larger for the high conservation farms.

A somewhat similar finding is reported for northwestern Wisconsin.[22] Here, the approach was to plan the most profitable cropping system that would meet certain soil erosion control standards which had been set by conservation specialists. Various alternative cropping systems were proposed that would yield higher net incomes than the typical existing systems, and that, in some cases, would also improve the conservation situation. Generally speaking, the more profitable systems involve more intensive operation, with higher costs of operation as well as higher incomes; and there is always the doubt whether these more intensive systems can be achieved with the levels of managerial skills available to farmers with presently lower incomes.

In northeastern Kansas, Charles C. Micheel and Charles W. Nauheim found that the availability of payments by the Agricultural Conservation Program (ACP) greatly affected the profitability of various soil conservation programs.[23] Without ACP payments, and

[21] Anderson, P. E. McNall, and Buis T. Inman, *Progress in Application of Soil-Conserving Practices, Southwestern Wisconsin*, ARS 43–44, Agricultural Research Service, U.S. Department of Agriculture, April 1957.

[22] John R. Schmidt and Rudolf A. Christiansen, *Potential Crop and Livestock Production and Net Farm Income on Dominant Soils in Northwest Wisconsin*, Research Bulletin 219, Agricultural Experiment Station, University of Wisconsin, Madison, Wis., May 1960.

[23] Micheel and Nauheim, *Economics of Soil Conservation, Northeastern Kansas*, Report 101, Kansas Agricultural Experiment Station, Manhattan, Kan., December 1961.

under present conditions, "the highest returns are produced by systems that include fertilizer but [are] without terraces"; if no arbitrary limit is placed on permissible soil losses, then systems without terraces are generally more profitable. Other systems can be devised that will hold soil losses to any desired level, but these usually are not profitable. The ACP payments greatly reduce the farmer's cost for specific soil conservation measures, and hence make it easier for him to undertake needed conservation practices. However, while ACP payments share the cost of capital of improvements, they do not help with annual maintenance costs. "Even though ACP payments reduce the cost of systems with terraces, the reduction is not great enough to make a system with terraces the most profitable on any soil under the present condition." One factor that may have strongly influenced these results is that present soil losses in the northeastern part of Kansas are rather moderate in comparison with severely eroding areas elsewhere—in the range of 5 tons to 9 tons per acre annually, even with the more erosive rotations.

Several studies of soil conservation in the series made at Iowa State College already have been discussed. All of these studies made by the agricultural college staff are concerned with profitability of soil conservation to some degree, but some deal more or less directly with this subject.[24] Heady and Allen conclude, in part:

> The greater output and income on high conservation farms was possible only through use of greater amounts of labor and capital. Farmers at the low end of the conservation scale were organized in the direction of cash grain farms, while those attaining the greatest degree of erosion control were in the direction of intensive livestock farms which used all of the farm-produced crops as feed. Low conservation farms might have increased income (especially at

[24] Among these, not already cited, are the following (all except the last listing being available from Iowa State University of Science and Technology, Ames, Iowa):
Earl O. Heady and Carl W. Allen, *Returns From and Capital Required for Soil Conservation Farming Systems*, Research Bulletin 381, May 1951.
Harald R. Jensen, Earl O. Heady, and Ross V. Baumann, *Costs, Returns and Capital Requirements for Soil-Conserving Farming on Rented Farms in Western Iowa*, Research Bulletin 423, March 1955.
Gerald W. Dean *et al.*, *Economic Optima in Soil Conservation Farming and Fertilizer Use for Farms in the Ida-Monona Soil Area of Western Iowa*, Research Bulletin 455, January 1958.
A. Gordon Ball, Earl O. Heady, and Ross V. Baumann, *Economic Evaluation of Use of Soil Conservation and Improvement Practices in Western Iowa*, Technical Bulletin 1162, U.S. Department of Agriculture, June 1957.

1945 and 1937–41 prices) simply by feeding all of the grain produced. While practices which would control erosion might be applied to the low conservation farms, income would not be as great as on the high conservation farms of the sample unless additional capital and labor were employed.

This study illustrates, perhaps as well as any, the fact that more intensive farm operation, with a greater demand for managerial competence as well as for more capital, may yield higher incomes and more soil conservation at the same time.

Jensen, Heady, and Baumann studied the operations of tenant-operated farms, to see the degree to which soil conservation practices were in the financial interest of both tenant and landlord. Their March 1955 report concludes, in part:

> The analysis shows: (a) The tenant's net income is increased by adjusting to soil conservation farming, irrespective of the leasing system; (b)The tenant's net income from a soil-conserving farming system including a dairy-hog program was larger under a crop-share-cash lease than under a livestock-share lease; (c) The landlord's net income was increased by adjusting to soil-conserving farming systems under a livestock-share lease (even after a commercial farm manager's fee had been paid) but not under a crop-share-cash lease. In other words, a landlord may realize less from a soil-conserving farming system if a crop-share lease is retained on the farm. The leasing system affected the average net income of the landlord.

In a further report on studies in the same general area, Baumann, Heady, and Aandahl considered numerous rotations and other ways in which soil erosion can be reduced to acceptable limits. Their June 1955 report stresses the need for more investment in livestock, buildings, and fences, as well as in terraces and annually in fertilizer.

> The change from present to alternative systems of farming would be profitable. . . . These larger net farm incomes for the soil-conserving systems would not be forthcoming immediately. For a year or two in the transition period, incomes would be lower than with present systems of farming. But as additional capital and labor are employed and become productive through yield-increasing rotations and other soil-management practices and through more livestock, net farm incomes would increase. And within a few years the accumulated net income from the conservation system would exceed the accumulated net income from present systems. The period in which this

would occur would be even shorter if allowance were made for the slow but continuous decline in yields of crops that will result if the soil-depleting system of grain farming now practiced is continued.[25]

In their further study of the same area, published in January 1958, Dean and his associates conclude:

The results of the study show that a combination of (1) rotations which include a maximum of corn within the range of rotations considered, (2) mechanical erosion-control practices (terraces, contouring and listing) and (3) high levels of fertilization provide the most profitable land-use program for most of the capital and resource situations studied. However, in instances where capital, labor or building space are not restricting resources, profits are maximized with a high-forage rotation. This type of rotational program allows maximum profits only at very high capital levels—where grain can be purchased and where the limit to cattle numbers is imposed by forage production. . . . While erosion control may be achieved either by mechanical practices or by high-forage rotations, greater farm profits generally are allowed by the former. . . . Farm incomes in the Ida-Monona soil area were drastically reduced by low livestock prices in the fall of 1955. . . . The major effect of changing hog prices, within the range considered, was on income rather than planning. Changes in feeding margins for beef cattle, however, required important shifts in farm plans for maximum profits.

Although these studies may have accurately revealed the profit potential in soil-conserving systems of farming, the studies by Held, Blase, and Timmons,[26] and the earlier studies cited therein, clearly show that a major proportion of all farmers in this area of Iowa do not now follow soil conservation systems, see no need to do so, and are opposed to the changes in farm organization that would be necessary to make such farming systems effective. The human resistance to change apparently is more serious than can be overcome by the budgeted increases in net farm income. Moreover, the difficulties of working out conservation programs on tenant-operated farms seem to be greater than the farm management researchers assumed.

[25] *Costs and Returns for Soil-Conserving Systems of Farming on Ida-Monona Soils in Iowa, op. cit.*
[26] *Soil Erosion and Some Means for Its Control, op. cit.*

Disassociation of Costs and Benefits

The intertemporal, interspace, and interpersonal comparisons, which frequently, if not invariably, are involved in soil conservation were discussed briefly in Chapter 1. As we pointed out there, these comparisons may arise in one of several ways. Not only is a large measure of social interrelationship nearly always required in connection with planning and carrying out soil conservation, but also basic physical interdependencies are involved. The actions of any person associated with the conservation program is likely to be affected by the actions of others, often in a complex pattern of interrelationships.

Some disassociation of costs and benefits takes place between geographic areas. The area of runoff in the watershed and the area of flood plain are largely distinct, for example. Also, the area from which silt or sediments are removed is distinct from the area in which they are deposited. Not only are the areas different, but they may be separated by relatively long distances. In limited cases, while the areas may be physically distinct they may be close enough together, and the units of landownership large enough, that the same landowner experiences runoff and flood, silt removal and deposition. Thus, soil materials might be washed from sloping land and deposited in the valley bottom on the same farm. But, more typically, the distances are large relative to the dimensions of the ownership units.

A disassociation of costs and benefits between different persons arises when the different kinds of physical areas are differently owned. Typically, one person bears the costs and another gets the benefits, or part of them. If there were no physical disassociations, there could be no personal disassociations. Some landowners are located in the flood plain, while others are located higher on the watershed, rarely experiencing floods on their land but perhaps contributing considerably to floods which occur elsewhere. Each landowner undertakes soil and water conservation measures which are rational in light of the costs and benefits which accrue to him. Society cannot reasonably expect a landowner to undertake soil or water conservation measures if the benefits accrue primarily to someone else. Society may, indeed, impose legal constraints or restraints on individual action, or may set up the legal machinery whereby one landowner or user may impose restraints against another; but, in practice, very little of this has been done in the United States.

These disassociations of costs and benefits are easier to see in the case of erosion and other damages caused by water, but they are sometimes present in damages caused by wind. Soil or sediments blown off one field or farm may damage other land downwind or create hazards, damage, or inconvenience to other persons located downwind. This, too, may happen far from the areas of origin. We have noted the extreme case when the dust storms of the mid-thirties blew soil from the Great Plains in observable amounts as far away as the Atlantic Seaboard.

Another form of disassociation arises between time periods. Investment in improvement of soil resources, or disinvestment of capital from soil, or new programs to change land use—all require time in order to be fully effective. At the very least, expected future costs and benefits must be discounted to a present net value, for comparative purposes, with all that this involves in terms of selecting the most appropriate rate of interest. But, as we have noted, time is also required before the farmer or other landowner adopts a new practice or program. Different landowners or farmers will move at different rates of speed, even when confronted with similar situations. Given the usual large differences in their personal situations, their rational responses will differ considerably.

Any form of disassociation of costs and benefits, or of program effects, necessarily constitutes a form of friction in the adoption of soil conservation programs. The seriousness of this friction will differ from one situation to another, but always on the side of constituting an obstacle or barrier, never on the side of facilitating ready adoption and operation of soil conservation measures. John F. Timmons has commented upon this general situation as follows:

> If all benefits and costs were perfectly associated for each participant, the problem of cost allocation would be relatively simple since participants receiving benefits would automatically bear the associated costs and participants bearing costs (or damages) would automatically receive compensating benefits. However, this perfect association of benefits and costs is far from being true due to several kinds of possible dissociations.
>
> There are two major factors responsible for dissociation of costs and benefits within the watershed. One factor consists of the shifting of benefits and costs between individuals within a particular time interval (intra- and inter-firm and intra-temporal). The second factor consists of the dissociation of costs and benefits between participants over time (intra- and inter-firm and inter-temporal).
>
> Referring to the first factor, an example of the intra-firm intra-

temporal dissociation of benefits and costs is a leasing arrangement which does not distribute benefits and costs equally between the two parties. An example of the inter-farm intra-temporal dissociation may be found in two adjoining farms in which one of the farmers operates his land in such a way as to cause serious flooding and siltation damages to the adjoining farm. Furthermore, the first mentioned farmer could not be expected to make substantial investments because part of the benefits therefrom would accrue to the other farmer.

Referring to the second factor, an example of intra-firm intertemporal dissociation would be a leasing arrangement of one year in duration without compensation provision. In such a case, the tenant would not be interested in making an investment in seedings, terraces or contour layouts since he would not be assured of getting full benefits from his investments. An example of inter-firm intertemporal dissociation would be exploitive farming systems in the upper section of a watershed resulting in the siltation over time of a city water reservoir or drainage ditches in the lower parts of the watershed.[27]

Institutional Framework

Modern man lives, works, and plays within a complex institutional framework created by the general society within which he lives. Laws, customs, and other relationships among persons create many specific arrangements or forces which influence, guide, or control individual action. The range of institutional arrangements is very great, and only a few of those with a direct bearing upon soil conservation will be considered here.

A basic institutional force in the United States is extensive private ownership of land. The landowner typically makes decisions as to land use, bears the costs associated with his decision, and reaps the benefits from it. Society may indeed limit the scope of individual decision in land use; arrangements differ greatly in different parts of the world. The United States has gone nearly to the end of the scale of institutional arrangements, in the direction of allowing the individual landowner the maximum freedom in use of his land. Land-use regulations of any kind are relatively uncommon in rural areas. There is a limited

[27] Timmons, "Economic Framework for Watershed Development," *Journal of Farm Economics,* Vol. 36, No. 5 (December 1954), pp. 1178–79.

amount of rural land zoning, more often in terms of settlement than of land use as such, and motivated as much by a desire to keep public costs to a minimum as by any other factor; but land-use regulations to govern conservation on farm land are most uncommon. As we have noted, these were once contemplated in the standard soil conservation districts act, but in practice they have been used rarely and have been repudiated when used. There are somewhat more land-use regulations in rural areas adjoining cities, but even there they are less common than within the cities. One individual is permitted to bring suit through the courts against another individual for losses arising out of the latter's actions on the land. This power is used, but not commonly, partly because its use is often costly relative to the possible gains. It has almost never been invoked to stop a use of land or practice which was aggravating floods, erosion (wind or water), or other undesirable effects upon any landowner or group of landowners. In the United States, the path of compulsion for soil conservation simply has not been chosen. This situation may change in the future, under the pressure of greater and more diverse demands for land.

The approach taken in the United States, with a set of institutions designed to make it effective, is the provision of public help in taking private action involving land use. The public help has taken the form of research, education, and direct subsidy. The research role of the U.S. Department of Agriculture and of the land-grant colleges is well-established. In any problem situation, the first cry is nearly always for more research, so that the facts shall be fully and accurately understood. The agricultural extension program is also well-established as a means of getting information to farmers and other landowners and users. In the last generation, the payment of direct cash subsidies to landowners and users has also become firmly established. No man is compelled to use his land in such a way as to reduce erosion or flood hazard to the minimum, but he will be given a subsidy as encouragement to do so. In the United States, the power of the public purse is vastly greater than such legal powers as police power, power of eminent domain, and the like. In the end, the power of the public purse rests on the power of taxation, for the government can distribute to some people only what it collects, often from others. The federal soil conservation programs—as expressed through the Soil Conservation Service (SCS), the Agricultural Stabilization and Conservation Service (ASCS), the Great Plains Conservation Program, and the various other special programs—all rest upon this broad group of institutional arrangements.

But society in the United States has also provided other institutional assistance to soil conservation. Notably, the soil conservation districts acts and other legislation which permit farmers and landowners to operate on a group basis for soil and water conservation ends are important. The role of the soil conservation districts may have proved to be less significant than their early enthusiastic supporters envisioned, but it has still been significant. Most individuals, including most farmers, respond to public opinion, especially to the opinion of the public with which they are best acquainted. The soil conservation districts have been an important means of forming and crystalizing public sentiment in favor of soil and water conservation, and they have undoubtedly carried along many farmers who might otherwise have been cool or hostile.

The foregoing institutional situations, by and large, facilitate soil conservation; other institutions are less helpful.

One of these is the system of land survey in the United States, and the resulting ownership boundaries; this arrangement has both facilitating and hindering aspects. A survey system which provides readily identifiable boundaries of property, and a system of land records and courts which makes the land titles enforceable, is certainly a help to individual action that takes account of the future as well as the present. In the absence of such clearly established boundaries and titles, land use necessarily would include a large measure of uncertainty, and society would be acting to shorten the planning horizon of the individual farmer. One need only look at the situation in a number of other countries, especially in Latin America, to realize how fortunate the United States is in this respect. At the same time, the rectangular survey system has generally resulted in farms where ownership boundaries, fence lines, field boundaries, and even rows within fields arbitrarily follow cardinal directions regardless of slope or other characteristics of the land. We have commented earlier upon the square lines in a basically rounded country, and on how the early SCS efforts were often directed at changing this situation. Many farmers have adopted contour farming, the most obvious and frequently most adequate cure, but not all farmers whose land might have been benefited thereby.

Another institutional arrangement that usually operates as a barrier to soil conservation is tenancy. There is much to be said in favor of farm tenancy; it is frequently the means whereby a farmer lacking capital may gain the use of a farm, or of a larger farm, and hence increase his income. But there have been many unsatisfactory aspects of tenancy in the United States, including the fact that most tenants have had relatively poor living conditions. As far as soil conservation is concerned, tenancy is a form of disassociation of costs and benefits. The landowner should make certain investments for soil conservation, yet he may not have a sufficient share in the returns to make his investment worthwhile; or the same may be true of the tenant. Almost without exception, tenancy shortens the planning horizon, since most tenancy arrangements do not provide any compensation to the tenant who moves for the value of any soil conservation improvements which he may have made. At the least, tenancy requires the concurrence of two decision makers, often with divergent interests; and thus the amount of soil and water conservation is almost invariably less than would have arisen from an owner-farmer.

Several specific studies have commented upon the adverse effects of farm tenancy upon soil conservation. In a study of share-rented farms in Texas, Calvin C. Boykin, Jr., in part, concluded:

> Insecurity of tenure was a major obstacle to conservation. The length of time a tenant expected to remain on the farm was more important in conservation accomplishments than the time he had been on the farm. Also, the presence of a livestock-share arrangement was conducive to greater conservation achievement. Cost-sharing between landowner and tenant was not done on permanent type conservation practices such as terracing and waterway development. Cost-sharing on temporary type practices such as grass seeding and cover cropping was done on most of the farms studied. Conservation accomplishments were greater on farms where costs were shared in proportion to returns. Results of the study indicate that farms owned by females made less progress in the district program than did farms owned by males. The age of tenants and landowners was not an important factor in conservation accomplishments.[28]

Haren commented as follows in his 1960 study of conservation farming in the Southern Piedmont:

> These and other changes in ownership and operation, including those accomplished before 1953, have enabled an increasing number of farmers to reduce intensity of cultivation and use effectively land

[28] Boykin, *Factors Affecting Conservation on Share-Rented Farms, Texas Blackland Prairie,* Progress Report 1879, Texas Agricultural Experiment Station, College Station, Tex., June 1956.

that formerly contributed little if anything to the farm income. Nevertheless, further changes in forthcoming years will be necessary in order to overcome such obstacles as those posed by (1) the small acreage in the majority of existing ownership units, (2) the dispersal of a relatively large number of holdings over two or more separate properties, (3) the continued reliance of many farmers on rented land, and (4) the prevalence of idle or semi-idle farms.[29]

In their study of conservation on selected farms of the Midwest, the North Central Farm Management Research Committee concluded, in part:

Tenure problems are one of the major "stumbling blocks" to the adoption of conservation practices in the Corn Belt. Making changes in farm organization, such as shifting land to hay and pasture and increasing livestock numbers to use the increased roughages produced, brings up questions on how to distribute costs and income.

Many farms in the Corn Belt are owned by absentee landlords who have little personal contact with their tenants. These owners do not realize that conservation adjustments will improve farm income over a period of several years. Instead, they want a high return on their investment *now*.

On many farms the tenant is also interested in short-run profits. He may have only a one-year lease with no assurance of renewal, or the leasing agreement may require him to shoulder a larger share of the conservation costs than he receives in benefits. He therefore plants a high proportion of the farm to cash crops—corn, soybeans and canning crops. Applying lime and fertilizer and increasing the acreage of legumes and grasses to conserve the soil may not increase current yields of grains sufficiently to offset the necessary reduction in grain acreage. Generally it is necessary to wait until the rotation has gone through a complete cycle and a crop of legumes has been plowed down on the entire farm before substantial increases in total grain production can be expected.[30]

Several of the Iowa State College studies specifically deal with tenancy as an obstacle to soil conservation. Frey discusses a number of

---

[29] *Op. cit.*

[30] *Conservation Problems and Achievements on Selected Midwestern Farms,* Special Circular 86, Ohio Agricultural Experiment Station, Wooster, Ohio, July 1951, p. 19.

problems arising out of tenancy.[31] He cites rental agreements covering only a year or a few years at most, the lack of co-operation from the landlord, inequitable arrangements in sharing costs and returns, and other factors. High fixed costs arising out of indebtedness were also a factor on owner-operated farms. Short expectancy of tenure was reported as a major obstacle by some tenants; and the small size of the farm, so that a reduction of row-crop acreage for soil conservation would leave an inadequate total crop acreage, was another adverse factor. These obstacles were serious enough when they occurred singly, but frequently two or more obstacles were present at the same time, thus making the situation much worse.

Jensen, Heady, and Baumann also considered the problems of tenant farms specifically in their 1955 report. While they felt that conservation farming was profitable for both landlord and tenant, they suggested several ways in which customary leasing arrangements might be improved:

> The modifications include: (a) Increase length of lease and security of tenure where feasible and in line with both tenant and landlord interests. (b) Encourage the tenant to invest in fertilizer and other semi-durable resources by including compensation provisions for portions of resources unexhausted upon termination of tenure on the farm. (c) Encourage investment in long-lived resources like buildings, terraces and tile by means of the landlord collecting improvement rent, or increasing his share of the product, or by sharing with the tenant the added costs and added returns. (d) Encourage optimum intensity of production in the short run by landlord-tenant sharing of variable costs in the same proportions as receipts and in the long run by tenant and landlord furnishing some of both fixed and variable resources and then sharing receipts in proportions similar to resources furnished. (e) Discourage cost transfers within the business by relating rental charges for the services of specialized resources directly to their productivity.[32]

A few years later, Held and Timmons stated, in their Research Bulletin 460:

> What was acceptable to an operator under a particular tenure situation, with given price and cost ratios, with a given financial situation and given objectives and with a given attitude toward the

[31] *Op. cit.*
[32] *Op. cit.*

problem of soil erosion, may be unworkable with changes in any or all of these factors. The major causes for failure to reduce soil losses during the period studied apparently were uncertainty of tenure, lack of adequate finances, greater reluctance to assume risk and lack of confidence in recommended practices.[33]

Fischer and Timmons, studying one soil conservation district, concluded:

The data obtained indicate that district progress was impeded significantly by (1) small size of farms, (2) tenant operatorship, (3) cash and crop-share leasing arrangements and (4) high inherent productivity of the land. . . . The following are reasons, beliefs or attitudes most often expressed by farm operators as contributing to their failure to follow district recommendations: (1) Insufficient cooperation between landlords and tenants in arranging for adoption and maintenance of recommended practices. (2) Belief that the practices were not necessary either because they would not adequately control erosion or because erosion was not excessive now. (3) Insufficient knowledge of the district's program and of the practices recommended. (4) Belief that application of recommended practices would increase capital and labor requirements without yielding commensurate additional income. (5) Farm and/or field layout would be such as to make recommended practices impractical. (6) Pressure of current financial obligations precluded the possibility of introducing practices which would increase current investment and/or reduce current income.[34]

Held, Blase, and Timmons, in their 1962 report, had this to say:

Land prices, taxes on land, interest rates, federal farm programs, and relative prices and costs determine the attractiveness of one type of farm enterprise as against others. In many instances in western Iowa, these factors tend to favor enterprises in which the erosion hazard is the greatest. One farmer, even a small group of farmers, can do little to change this economic and institutional environment. The general public should be aware of the repercussions that policies, set for accomplishing other objectives, may have with respect to erosion control.[35]

[33] *Op. cit.*
[34] *Op. cit.*
[35] *Op. cit.*

Government Programs: Aid or
Hindrance to Soil Conservation?

Partly in recognition of these hindrances to soil conservation, and partly for other reasons, the various government—primarily federal—programs for soil conservation have been enacted. Perhaps more basic than these governmental programs has been a public or social interest in natural resources generally, and more particularly in soil resources. As we noted early in this book, many Americans are more nearly neo-Malthusians than is generally realized. To a large extent, public attitudes toward resources are expressed through the price system. Not only are prices bid up for those resources which are now scarce or highly productive, or both; but also they are bid up for those resources which are reasonably expected to be relatively scarce in the future. Higher prices both ration scarce resources to the more productive uses and provide a signal to the whole economy and society. No responsible economist argues that the price system works perfectly; divergence of views arises as to whether better performance can be secured by overt intervention than by trying to improve its functioning. The price system has taken very little note of soil conservation, as we have noted, either before 1933 when there was little national concern about it or in more recent times, in the sense of offering higher prices for land in better conservation condition.

There has been much sentiment, in professional as well as in popular circles, for a greater degree of public concern over soil conservation than has shown up in the price system. It can be argued that the price system works badly, because of poor information or for other reasons; or that society must take a longer view than that of the individuals whose collective action determines resource prices. And still other grounds can be advanced for a larger public role. We shall not here consider the pros and cons of these arguments at any length, but simply note that substantial public support has existed for soil conservation programs beyond those of the free price system.

The various federal and other governmental programs directed at soil conservation have almost certainly helped to bring about a larger degree of soil conservation than would otherwise have existed. The general public, and farmers in particular, have been aroused to the hazards of soil erosion, have been educated as to means for its control, and have been cajoled and encouraged with financial aid to undertake specific soil conservation measures. Certainly, much of the identifiable soil conservation activity on the land has been associated with one or more of the governmental programs for soil conservation. A major share of the terraces built, for instance, have had some federal financial aid. One cannot say with certainty that none of these would have been built without such aid, but without doubt the number would have been much less.

At the same time, the public (especially federal) programs to help soil conservation have had some serious weaknesses, which have greatly impeded the amount of soil conservation achieved from the public funds expended. A balanced appraisal of soil conservation must take the more serious of these weaknesses into account; and we shall briefly consider them here. Since the very real achievements of government programs have been examined previously, however, we shall not repeat the discussion.

The federal soil conservation programs have suffered from three major inconsistencies: (1) the incompatibility between public programs to raise farm incomes through the medium of soil conservation payments and the achievement of actual soil conservation; (2) the incompatibility between the need of the Soil Conservation Service to build popular and political support everywhere it could, and the soil conservation needs in various parts of the country; and (3) the basic incompatibility between measures designed to increase agricultural output, beyond those required merely to maintain basic productive capacity, and other more expensive programs for control of agricultural output.

There can be little question that when the Soil Conservation and Domestic Allotment Act was passed in 1936, after the original Agricultural Adjustment Act was declared unconstitutional, it was done as a means of continuing direct income payments to farmers. The basic purpose seems clearly to have been income support or augmentation, in spite of protestations to the contrary. Yet farmers were required to "earn" their payments by "soil conservation" measures. Substantial sums of money were spent for this program, and some soil conservation accomplishments were surely secured. Yet the conclusion seems inescapable that the resulting amount of soil conservation was limited in relation to the expenditure, and limited in no small part because of the fundamental incompatibility of these two purposes of the legislation. It is easy to see in Figures 27, 28, and 29 that payments under this program have been more closely related to the number of farms than to any measure of soil conservation "need."

The bureaucratic and political struggles of SCS have led to the second incompatibility. The agency was born in conflict, survived only by continuous fighting, and is secure today only because it mends its political fences continuously. By "political" we do not mean partisan, for the agency has avoided almost entirely an identification with either major national political party, seeking instead support within each. We do mean a broad base of public support, especially among farmers, which will lead to more nearly "adequate" appropriations, to helpful legislation, and to stopping attempted raids from any source. In this sense, every public agency is to a degree political; but the Soil Conservation Service has had to be more politically conscious than most in view of its numerous and powerful enemies—notably the federal and state agricultural extension services, the predecessors of the Agricultural Stabilization and Conservation Service, and others. In order to win broad support, SCS had to have a program which could appeal to a considerable proportion of all farmers in every major region and, if possible, in each congressional district. In those areas where its own earlier surveys had shown no major soil erosion problems, it had to devise other appealing approaches. As we noted in Chapter 4, the content of the SCS program has changed greatly. It has evolved from major emphasis upon erosion control or the maintenance of basic productive capacity to a more complex program of land management involving construction and new basic productive capacity and stimulation of greater annual inputs as well as maintenance of basic capacity. If land leveling, or drainage, or reseeding range land, or any one of various other land management practices could be made popular in an area, SCS was under strong temptation to adopt the practice—even if its value for soil erosion control was limited or absent. In a great many instances, it has adopted programs of this kind. This can be defended, on the one hand, as commendable flexibility on the part of a public agency to meet the needs of its constituent public. Or it can be criticized, on the other hand, as politically motivated expediency. Much depends on the viewpoint of the observer.

Adoption of such programs led directly into the third basic incompatibility, that between output-increasing and output-controlling programs. Many of the programs undertaken in the name of soil conservation, by SCS or ASCS and its predecessors, have added substantially to agricultural output, either through stimulating increased annual inputs or by building additional basic productive capacity which would increase output in future years, or both. As we have noted previously, the federal soil conservation programs were not the only, nor perhaps the chief, federal and state activities stimulating agricultural output. At the same time, other parts of the U.S. Depart-

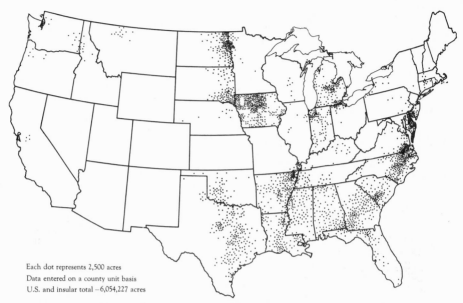

Each dot represents 2,500 acres
Data entered on a county unit basis
U.S. and insular total —6,054,227 acres

Figure 42. Establishment of green manure and cover crops, 1963. (Map from Agricultural Stabilization and Conservation Service, U.S. Department of Agriculture.)

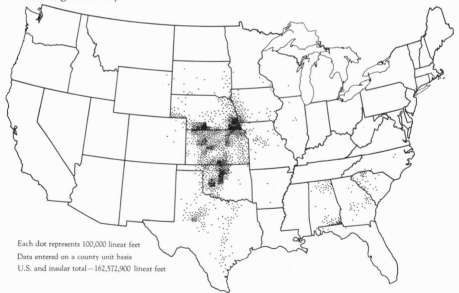

Each dot represents 100,000 linear feet
Data entered on a county unit basis
U.S. and insular total—162,572,900 linear feet

Figure 43. Standard terraces to control erosion and runoff, 1963. (Map from Agricultural Stabilization and Conservation Service, U.S. Department of Agriculture.)

ment of Agriculture were spending much larger sums in trying to control agricultural output, to store unwanted surpluses, to maintain prices at a higher level than the market would support, to divert surpluses into stimulated domestic consumption channels, to push exports abroad on much more generous terms than for strictly commercial trade transactions, or in some combination of these programs. Given the divergent political forces or supporting groups within agriculture, the Department of Agriculture may have had little alternative than to act as it has, but the incompatibility of its actions is nonetheless serious. The totality of all public (federal and state) programs is still less consistent internally.

Statistics which report the accomplishment of erosion control practices on farms indicate that there have been significantly different responses on a geographical basis to the efforts made to secure the adoption of such measures. (See, for example, Figures 42, 43, and 44.) This is understandable, in part, simply because the soil erosion hazard is not uniformly serious throughout the nation. Furthermore, differences in types of farming make it relatively simple to accomplish certain measures in one area but more difficult to do so elsewhere. The dairy farmer who can use large quantities of hay and pasture is in a better position to control erosion on slopes through the use of sod crops

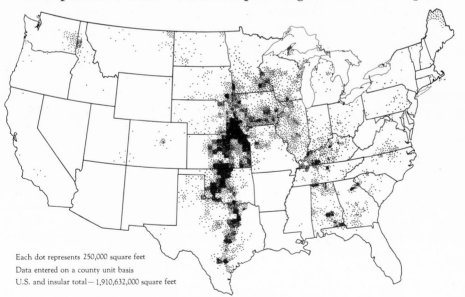

Each dot represents 250,000 square feet
Data entered on a county unit basis
U.S. and insular total— 1,910,632,000 square feet

Figure 44. Establishment of permanent sod waterways, 1961. (Map from Agricultural Stabilization and Conservation Service, U.S. Department of Agriculture.)

than a farmer whose farming enterprise is built around row crops. Where important investments are required, either to put practices into effect or to make the subsequent adjustments in farming operations that an erosion control plan might require, the differences in the economic health of farm people in various parts of the country may be an important factor in explaining some of these differences. Not to be forgotten either is the fact that problems of a Texas cotton farmer are not the same as those for a cotton farmer in the Mississippi Delta or in the southeastern part of the Piedmont. The cash-grain farmer of the Corn Belt does not have the same problems as the livestock farmer in the same area; and neither has the same problems as the wheat farmer on the Great Plains, the dairy farmer in the Lake States, or the dairy farmer in the Northeast. Variations in climate, soils, topography and methods of farming limit the practical use of some measures to certain areas. Other practices may substitute for one another, although they may not necessarily accomplish the same degree of erosion control.

With these limitations in mind, the dot maps shown in Figure 30 through Figure 35, which were based upon data from the 1959 Census of Agriculture, and Figure 42 through Figure 44 showing ACP accomplishments for 1959 are useful in demonstrating the unevenness of the farmers' response to these efforts. (It should be noted, however, that in the ACP maps activity for one year only is shown, and because of the relative permanence of terraces and sod waterways, a truer picture requires the use of a map such as Figure 30 shows.)

The establishment of permanent sod waterways is one of the lesser used practices, and because in Figure 44 each dot represents very few acres compared to those indicated by dots in the other maps, it is not really comparable. However, it is interesting because it shows rather clearly the striking correspondence between political boundaries and the concentration of effort or the lack of effort. For example, the principal effort made in both 1961 and in previous years was in Kansas and in the eastern two-thirds of that state. The western third of the state is largely untouched, as is the greater part of the nation westward of that area—an understandable situation because of the drier climate there. However, this does not explain why the use of sod waterways stops so abruptly at the border between Kansas and Missouri, or at the southeastern border of Kansas with Oklahoma. There are also interesting islands of white in Kansas itself. While one must allow for physical differences, these are seldom so great that what is a serious problem in one jurisdiction is suddenly no problem at all in the adjoining one. County boundaries, particularly noticeable in the case of the counties of the blackland prairie area of Texas, and state boundaries can be readily established because of the relative autonomy which is permitted the county and state committees, the links between the ACP and the farmer.

The ACP effort receives only broad, general direction at the federal level. Here, certain policies are established, and a docket, or list of practices approved for federal cost-sharing is drawn up. Each state in turn prepares, in cafeteria fashion, a docket of its own, making its selections from the federal docket. The state committee is under no particular pressure to include specific practices on its docket or to give them special emphasis; these decisions are solely within the committee's province. In turn, the state docket is only a guide for the use of the county committees in building their dockets and is a permissive, rather than a required, list of practices.

Additional flexibility is provided by permitting the state committees to determine, within certain limits, the federal share of the cost for each practice, to establish the minimum standards of performance, to adapt certain practices to fit conditions in their states, and with the approval of the program administrator to include in either the state docket or the docket of one or more counties certain practices not listed in the federal docket.

Whether or not practices are ultimately carried out lies with the farmers in any county, but the state and county committees are not without influence in these matters. The funds available for this program are relatively fixed at both state and county levels. If certain practices receive more attention from farmers than other practices which a committee feels are more important, the committee has the option of deleting some of the competing practices from its docket in order to encourage farmers to move in a particular direction. Certainly, the authority to determine the federal share of cost for the practices provides another way in which the state or county committee can emphasize particular measures which it considers desirable. That not all committees respond in the same way to a situation in which the physical factors may be much the same is the conclusion one must reach when attempting to understand the differences in program adoption which apparently result from differences in county or state committee policies. This is but one way in which the administration of programs influences the performance. An important measure of credit in some instances, and criticism in other instances, must be given to the county committees and the support or leadership given them by the state committees.

Because we are concerned with the variations in the response of farmers to a number of programs, we have looked for something which might provide us with an index of the degree of program acceptance in different areas. A variety of practices of varying degrees of effectiveness and practices which are not always widely applicable must somehow be weighted, totaled, and related to the particular physical conditions of a specific site. One reasonably good index of all program adoption is provided by the extent to which contouring has been used in an area.

There are several reasons for choosing contouring for our index. First of all, it is a practice that is widely applicable and actually seems to be used on more acreage throughout the nation than any other practice of comparable effectiveness. It is difficult to know just what the extent of certain practices may be at any point in time, for SCS generally reports only those practices which have been newly established with their assistance. The statistics by ACP provide a cumulative report on practices for which they have shared costs, but this reporting includes only the practices for which farmers received financial assistance. There is also the possibility in some instances that the acreage of the practice may have been counted more than once. The Census of Agriculture for 1959, however, overcame these difficulties with a report of all acreage on which four different practices, including contouring, were being used in that year.

Contouring is a relatively inexpensive practice, unless large-scale rebuilding of fences at field boundaries is required. It is a comparatively simple practice to follow, once the contour lines have been established. Its effectiveness in controlling erosion is considerable under some circumstances, for it has been shown on an experimental basis that erosion losses may be cut by as much as one-half after contouring is substituted for the conventional up-and-down-hill type of farming operations.

It seems reasonable to expect that farmers who accept contouring when its use would be feasible would be likely to accept and use other erosion control measures, whereas a farmer who did not do so could be expected to use other erosion control practices only to a limited extent. The exceptions here would be those instances where strip cropping tended to be a substitute for contouring and in those situations where terraces had been installed. The Census enumeration of the practice of contouring apparently excluded from its definition of contouring that done almost automatically after terraces are installed.

The adequacy of contouring, strip cropping, and terracing as indicators of a reasonably effective job of erosion control can be given a rough check by comparing acreage figures available from the 1959 Census of Agriculture and the 1957 National Inventory of Soil and Water Conservation Needs (CNI). The Census reports for these practices some 70.3 million acres; the CNI indicates that some 62.5 million acres of cropland were either adequately treated for the erosion hazard to which they were subject or were not feasible to treat. The 70.3 million acres, it should be noted, include pastureland as well as cropland. The proportion of the cropland on which terraces had been constructed ranged from less than 10 per cent in five subregions to over 40 per cent in one of eleven subregions analyzed (Figure 45).

The programs for production controls and land-use adjustment of recent years have been a factor of undetermined importance in the reduction of soil losses. In some areas, particularly the Southeast, they seem to have served as a catalyst in bringing about shifts in land use that have reduced the erosion hazard. The potential of these programs has never been fully realized, however, as illustrated by the experience of the Conservation Reserve Program of the Soil Bank of the ASCS (for which authority to contract for retirement of cropland for tem-

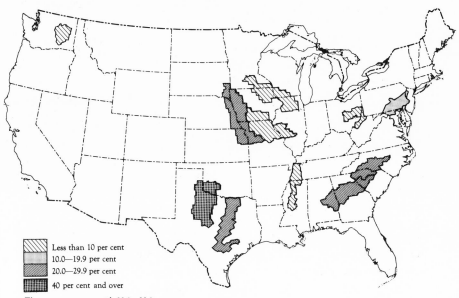

Less than 10 per cent
10.0—19.9 per cent
20.0—29.9 per cent
40 per cent and over

There were no areas with 30.0—39.9 per cent

Figure 45. Area contoured as percentage of cropland harvested in 11 selected economic areas, 1959.

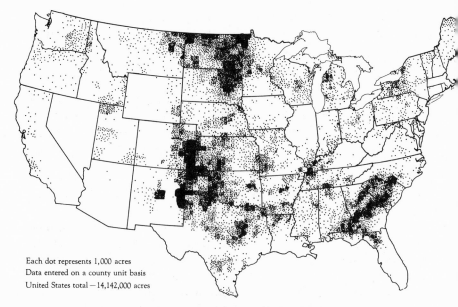

Each dot represents 1,000 acres
Data entered on a county unit basis
United States total — 14,142,000 acres

Figure 46. Cropland in Conservation Reserve Program, acreage under contract on January 1, 1965. (Map from Agricultural Stabilization and Conservation Service, U.S. Department of Agriculture.)

porary transfer to conservation uses expired in 1960). The amount of cropland under contract in that program on January 1, 1965 is shown in Figure 46 and the cumulative totals and net changes in acreage under contract are indicated in Figure 47. Participation in the Conservation Reserve Program was on a voluntary basis; there was no deliberate effort to make it more attractive to farm operators who were cropping land with a high erosion hazard so as to induce them to participate. As a result, land with a relatively small erosion hazard could have been retired from production in some areas while land with a severe erosion hazard remained in production in other areas. Studies conducted in different sections of the nation suggest, however, that the quality of land which was put into the conservation reserve in a particular area was generally, but not always, of lower than average quality for that area.[36] However, it was not only the less productive land that was retired. In some instances, the farm operator was elderly

[36] Raymond P. Christensen and Ronald O. Aires, *Economic Effects of Acreage Control Programs in the 1950's*, Agricultural Economic Report 18, Economic Research Service, U.S. Department of Agriculture, October 1962.

and saw the program as an opportunity to reduce the size of his operations, or to retire. In others, participation was greatest among farmers who had nonfarm jobs.

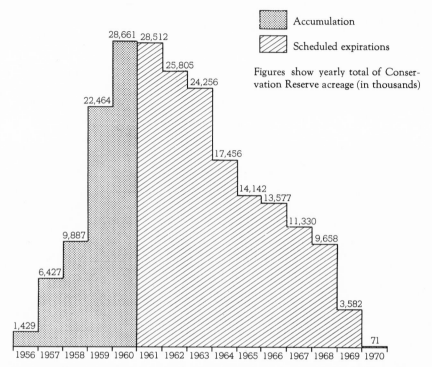

Figure 47. Conservation Reserve Program, total acreage and net change by years, 1956–70. (Based on chart from Agricultural Stabilization and Conservation Service, U.S. Department of Agriculture.)

PART III

THE NEXT GENERATION

Chapter 11

THE FUTURE ECONOMY
AND SOCIETY

PART I of this book was a historical review of the evolution of soil conservation programs, public and private, in the United States; Part II was an inventory, a stock-taking, of where the United States is today so far as soil conservation is concerned. This examination of the country's past and present resource effort forms the basis for our speculations about the future directions of soil conservation. Looking ahead is always difficult. No one can know what will happen in the future, and perhaps it is better so. But we try in our last two chapters to make explicit what is implicit in our analysis of the past and present.

Chapter 11 presents a brief description of the economy and society of the future, as these now look. In this chapter, we summarize what others have written, deliberately adding only a few additional points of our own. Times have changed greatly since Bennett and his co-workers first established a national program of soil conservation. The soil conservation program of the federal government has undergone major changes in the ensuing generation, although all the program changes that an evolving social and economic environment might bring about perhaps have not yet been achieved. If the total social and economic framework develops during the next generation as we think it will, then further changes in the soil conservation programs are surely called for. We close this book in Chapter 12, by trying to suggest some of the changes we expect. No one can possibly foresee every change or development, and we make no claim for comprehensiveness nor omniscience; however, we can raise at least some of the pertinent issues and suggest at least some of the major alternatives for the future.

It is useful in this connection to distinguish between forecasts and projections. Projections, on the one hand, are estimates of future events, based upon certain assumptions. If the assumptions are sufficient, in the mathematician's sense of the term, the projections flow naturally and any analyst would reach the same conclusions. The reasonableness of the results rests on the reasonableness of the initial assumptions and upon the coefficients or interrelationships between variables and results. A forecast, on the other hand, is a person's estimate of what is most likely to happen. While it, too, is based upon assumptions and interrelations, there need not be a simple or mathematical relation between these bases and the results. The forecaster may introduce other factors, either deliberately or unconsciously.

In these two concluding chapters, we are projecting, not forecasting; yet we choose assumptions for their reasonableness. Our basic assumption, so far as soil conservation is concerned, is that, in the future as in the past, both public and private soil conservation efforts will be shaped more by broad changes in the total society and economy than by anything that may be done as part of soil conservation programs.

General Shape of the Economy and Society[1]

Traditionally, the first place to start in looking toward the future economy and society has been a projection of total population. Barring a catastrophic war, the population of the United States will increase over the decades ahead. The only real question is: how much? A reasonable projection of population, neither extremely high nor extremely low, is 245 million people in 1980 and 330 million in the year 2000. There is a consensus that most of the net increase will be in suburban areas around large and moderate-sized city or metropolitan areas. City centers have lost population in recent years, partly because activities other than for residential purposes have taken over increasing acreage in city centers, but also partly because most people prefer suburban to city living. Some evidences of a reverse movement exist, but the magnitude of the future urban population change is so great that a major share of the increased population necessarily will be located in the suburbs. Farm population has declined, and will surely continue to decline further. While nonfarm population in true open

---

[1] Many projections and forecasts have been made of the factors here considered; a full review of such studies would be a major project in itself. We lean heavily on Hans H. Landsberg, Leonard L. Fischman, and Joseph L. Fisher, *Resources in America's Future: Patterns of Requirements and Availabilities, 1960–2000* (Baltimore: The Johns Hopkins Press for Resources for the Future, Inc., 1963).

country has increased in the past and is likely to increase in the future, this may not much more than offset the decline in farm population.

There is also a consensus among economists that total national product and per capita income will rise in the future—always on the assumption of no catastrophic war. A medium projection is for a total gross national product of $1,060 billion in 1980 and $2,200 billion in 2000, each contrasted with $504 billion in 1960. In other words, GNP will, at the very least, double in each twenty years. Somewhat lower and considerably higher figures could reasonably be projected, if either more pessimistic or optimistic assumptions were made. Some variations in the use of gross national product are possible, but the greater part of it will almost surely go for personal expenditures of one kind or another. A reasonable estimate is that expenditures for personal consumption per capita (in terms of 1960 prices) will rise from $1,830 in 1960 to $2,700 in 1980 and $4,000 by 2000. In other words, during each 20-year period, per capita expenditures will increase by about 50 per cent, compared with 100 per cent in total output, the difference being accounted for by the increased total population.

For as long as accurate statistics have been available, two basic trends in the use of all kinds of natural resources have been evident: (1) an increasing per capita consumption as gross output per person has risen; (2) a declining resource input per unit of total gross output. The latter relationship, in turn, has been caused by two basic forces: (a) an increasing degree of processing of each unit, so that more total value is created with each unit of natural resources; and (b) a greater relative increase in the role of services in the economy, so that manufactures and other uses of natural resources have increased less rapidly. It seems probable that these same general trends will continue well into the future.

One other factor of a general nature requires mention at this point. Total leisure has increased in the United States, and seems likely to increase further. If one defines leisure as all time not used in work, sleep, and necessary personal chores, then from 177 billion hours for the whole nation in 1900 it has increased to 453 billion hours in 1950, and a reasonable projection is 1,113 billion hours in 2000.[2] The amount of leisure time depends not merely upon the average working hours of those in the labor force, but also upon the relative numbers of young people who have not reached working age, the number of older retired

[2] Marion Clawson, *Land and Water for Recreation* (Chicago: Rand McNally & Co., 1963).

persons, and the number of middle-aged persons not working for one reason or another. Leisure time also depends upon the number of persons enjoying annual vacations, and their average length, as well as upon the length of the typical workweek. Some leisure is daily, largely after work or after school; some is weekly, usually week end; and some is annual, primarily paid vacation. Each form bids fair to increase over the next several decades.

Demand for Products from the Land

These estimates of the total economy and society can be broken down to apply more specifically to soil conservation by estimates of the demand for various products or services of the land.[3] The gross value of farm products sold from farms, measured at the farm gate, rose from $20 billion (in 1960 dollars) in 1940 to $29 billion in 1960, and is projected to rise to $42 billion in 1980 and to $57 billion in 2000. These increases, while substantial, are materially less than those increases projected in gross national product; as the nation's output rises, a smaller relative share of it comes from agriculture. This can be expressed in another way; consumers' expenditures for food and tobacco, at retail in 1960, were $487 per capita (in 1960 prices), or nearly 26 per cent of total consumer expenditures; by 1980, per capita expenditures in terms of the same prices will rise to $600, but as a percentage of total expenditures will fall to 22 per cent; and, by 2000, they will rise to $726, but fall further to 18 per cent of the total. Even these amounts and percentages will be heavily influenced by extensive marketing and processing services and costs. The quantity of food required will rise in proportion to the increase in total population; but higher per capita incomes will have very little effect upon demand for food, especially at the farm.

Two agricultural commodities—wheat and meat—are of particular

[3] There is an extensive literature on this general subject, the summarizing of which would be a major project. For two recent reports, see Earl O. Heady and Luther G. Tweeten, *Resource Demand and Structure of the Agricultural Industry,* Iowa State University Press, Ames, Iowa, 1963; and James T. Bonnen, "Analyses of Long-run Economic Projections for American Agriculture," *Agricultural Research in the Space Age,* Proceedings of the Twelfth Annual Meeting of the Agricultural Research Institute (Washington: National Academy of Sciences–National Research Council, 1963), and references cited therein. We use primarily the material in Landsberg, Fischman, and Fisher, *op. cit.*

interest for soil conservation, partly because each uses relatively large areas of land, and partly because one form of soil conservation would be to shift some land from one to the other use. The quantity of wheat eaten by the American consumer rose from 14 million tons (grain equivalent) in 1920 to only 15 million tons in 1960; reasonable projections are 16.5 million tons by 1980 and 20 million tons by 2000. However, it would not be unreasonable to estimate that total wheat consumption by humans in the United States will decrease, rather than increase, in the future. As incomes rise, people tend to buy more of other foods and less of bread—the latter is an inferior good, in an economic sense, because its consumption is inversely correlated with income, at least after incomes pass some point. Population increases will not mean large, or perhaps any, increase in domestic demand for wheat; and if demand stays constant, or nearly so, acreage required to produce for the domestic market will almost certainly decline if yields continue their historic upward march. Foreign demand for American wheat is a complex matter, as much political as economic. If the United States is willing to give wheat to low-income countries, there is a virtually unlimited market; but both political and economic considerations may well limit such gifts. Sales to Soviet Russia and other iron curtain countries are a major unknown.

In contrast, the consumption of meat has risen from 7 million tons (carcass weight) in 1920 to 14.5 million tons in 1960, and is projected to rise to 23 million tons in 1980 and 32.5 million tons in 2000. Of this, beef has been a rising share and pork a declining share in recent years. As per capita incomes rise, beef consumption tends to rise more than does pork consumption. The increased beef required to meet this consumption will come from many sources, but some surely can, and should, come from more extensive and improved grasslands. One obvious shift in land use, and one that would contribute importantly to better soil conservation, would be from wheat to grass; but both prices and programs are great obstacles to such shifts.[4]

The demand for forest products generally will increase greatly in the future. Pulp consumption is likely to move upward from 27 million tons in 1960 to 56 million tons in 1980 and to 110 million tons in 2000. The American society in many ways is a "paper society"; paper is critical to the flow of information and business, and increasingly is the

[4] Roger H. Willsie, *The Economics of Classifying Farmland between Alternative Uses,* Research Bulletin 208, Nebraska Agricultural Experiment Station, Lincoln, Neb., March 1963.

dominant form of packaging. The consumption of plywood is likely to increase even more on a relative scale than the figures estimated for pulp consumption, although on an absolute basis the increase will be somewhat less. Plywood is coming increasingly to be used during the construction process, as well as for parts of buildings, furniture, and other consumption goods. If the price of lumber were to stay constant, and good quality were readily available, the potential demand could push consumption upward materially from 1960 levels. Although rising prices may well tend to hold down consumption, some increase in total consumption seems highly probable. All of this means increased demand for forest products in general; yet the total forest area, for a combination of reasons, is not likely to increase and may even decrease.

Outdoor recreation is another use of land which has been mounting rapidly in recent years, and various projections of future demand have been made.[5] Use of land for recreation has doubled or more since 1956, and our own estimate is for a tenfold increase from 1956 to 2000.[6] Other estimates have been lower. The increases in demand are likely to be relatively greatest for the areas that are more distant and inherently more attractive for recreational uses; increased per capita income and improved travel facilities are likely to be most effective for such areas. But only moderately attractive areas, if conveniently located with respect to where considerable numbers of people live, also will be in high demand. There are many serious organizational and financial problems involved in transforming farmland into recreation land, and the scale of such transformation cannot possibly solve the problems of excess agricultural production with which the country is now laboring. But this is an important shift in land use; moreover, it is one which often leads to better conservation practices on the land.

The use of water for various purposes has been growing and will continue to do so; along with this further demand, watershed management will become an increasingly important purpose of land management. It will be necessary to make clearer distinctions as to withdrawal, use, and depletion of water. Some water is withdrawn from streams and wells, but this does not mean the same amount of water necessarily is used. The use of water may be greater than the amount withdrawn, for example, if the same water is re-circulated; however,

[5] Outdoor Recreation Resources Review Commission, *Outdoor Recreation for America, a Report to the President and to the Congress* (Washington: Government Printing Office, 1962).

[6] Marion Clawson, R. Burnell Held, and C. H. Stoddard, *Land for the Future* (Baltimore: The Johns Hopkins Press for Resources for the Future, Inc., 1960).

the use could be equal to withdrawal or, in some cases, even less. Then some water is used up, through depletion—evaporation, seepage to underground strata, transpired, and so on. Each of these measures is important, and their differences may be exceptionally important in some situations. The trend generally in each has been, and will be, upward. Use will probably nearly double from 1960 to 2000—more for some purposes, less for others. The problem is partly one of the quantity of water, but more particularly it is one of quality. Nationally, there is no shortage of water, but in the arid regions the total water supply may be so limited that it will be necessary to curb some water-using activities. Water pollution is a problem everywhere. Since one major pollutant is silt and sediment from eroding land, soil conservation measures are important as a means of reducing pollution. More land is likely to be managed in the future primarily for watershed purposes, even at the expense of other output from the land, than in the past.

The complex urban-dominated society of the future also will make a number of other demands on land—demands that in the past might have been considered unusual. Each of these will take relatively small areas, but in the aggregate will use a considerable amount of land. Included among such areas are industrial sites, urban transport facilities, rural residences, vacation homes, and even waste disposal areas. Park and recreation areas, or only open spaces near cities, will require strategic sites.

Role of Science and Technology

Science and technology have been powerful factors affecting land use in the past, and there is every reason to believe that they will continue to be important. Their effect is most obvious and most easily measurable for agriculture, as indicated by the data showing the large rise in agricultural output from a generally decreasing total input into agriculture. But almost every other land use has felt their effects also.

The United States has a vast complex of institutions contributing to the supply of knowledge and influencing resource use generally. Several departments of the federal government, the land-grant colleges and other colleges, universities, and numerous private institutions operate in this way. By and large, the results of their research have been output-increasing in agriculture and in other land use. The financial support and other forms of assistance for these varied institutions generally has been rising and may be expected to continue upward; as a result a continued flow of research results may be expected. In this book, and more particularly as we look ahead, a continued stream of new research results is assumed. We cannot, nor need we, specify exactly what these results will be; we simply assume that new research will continue to flow and be effective. We do not fully understand the processes by which new scientific knowledge has been translated into practical agricultural operations in the past, but we can observe its effects. Similarly, we may not be sure exactly how it will operate in the future, but assume that new knowledge will in fact be translated into new and more efficient practices as it becomes available. At any given date, there is a considerable backlog of un-applied, or at least partially applied, knowledge; as time goes on, some of this will be translated into current practice while new knowledge is becoming available.

New scientific knowledge and the new application of existing but unapplied knowledge will increase the output of most products from the land. Thus, farms can be expected to produce a greater output of crops from a smaller total input of all factors in the future than today; and the same may well be true for forestry, grazing, recreation, and perhaps other major land uses as well. In the case of recreation, this will largely consist of a more intensive rate of use; and there is grave danger that this will lower the quality of the recreation experience.

The question is not so much whether output will increase—we think it will for most land uses—but how much it will rise. Bonnen has pointed out how almost every serious effort at long-range projection of agricultural supply and demand has underestimated the rate of in-crease in agricultural output in relation to agriculture inputs.[7] Produc-tion specialists who worry over current difficulties and problems often have had a poor record of forecasting future technological change; their current problems have made them too pessimistic. For the pur-poses of this book, it is not necessary to make specific quantitative fore-casts or projections of future agricultural output from a specified level of inputs. We can simply say that the output will rise in relation to input, and that agricultural supplies will be ample in relation to demand.

The Number, Size, and Output of Farms

A combination of economic, social, and demographic factors will cer-tainly lead to further major reductions in the total number of farms during the next generation; the only real question is just how far the process will go.

[7] *Op. cit.*

Young men, especially the sons of farmers, actively sought to enter farming in the past. A generation or so ago, much was made of the "agricultural ladder"; the farmer's son was supposed to begin as a hired man, save his money, rent a farm, save his money, and ultimately buy a farm, burdened at first with a mortgage which in time got paid off. This route was followed by many men; but perhaps, at all stages, as many as—possibly more than—those who made it failed to advance up the ladder. And some young men never were forced through these various stages but inherited a farm and began as farmers. There is less emphasis upon the agricultural ladder today, for several reasons. The hired farm man of bygone times has all but disappeared; the total of hired labor is less today than a generation ago, and most of today's hired labor force is permanently cast in that role. Farm tenancy has changed materially; there are fewer full tenants, but renting of additional land to supplement a farm of inadequate size has become far more common. The purchase of a farm by a young man with limited capital is much more difficult today because of inflated land prices in relation to income from farming.

The farmer's young son of today is much more likely to consider an occupation other than farming as his alternative to farm operation. Young men and farmers up to an age of perhaps 35 years are either uncommitted to farming as a way of life, or can more readily withdraw for another job, perhaps an urban one. After some point, however, such shifts become very difficult—economically, sociologically, and psychologically. The men may not know where to find another job; in seeking other work, they might suffer severe discrimination because of age, or race, or lack of job skills; and they would face a serious readjustment problem if they got a job. Under these circumstances, such persons stay with farming even though incomes are low.

The big decline in the total number of farms—down in 1964 to half, or possibly less, of their one-time peak—is marked by the large drop in the number of young men who have entered farming, not in the accelerated withdrawal of older men from farming.[8] Depending upon the assumptions made, Don Kanel estimates that as much as 25 per cent of the total decline in the number of farms might have resulted from the accelerated withdrawal of older farmers, whereas the

[8] See Chapter 5 and the following articles in the February 1963 issue (Vol. 45, No. 1) of the *Journal of Farm Economics:* Jackson V. McElveen, "Farm Numbers, Farm Size and Farm Income"; Marion Clawson, "Aging Farmers and Agricultural Policy"; G. S. Tolley and H. W. Hjort, "Age-Mobility and Southern Farmer Skill—Looking Ahead for Area Development"; and Don Kanel, "Farm Adjustment by Age Groups, North Central States, 1950–1959."

Clawson estimate indicates no more than 8 per cent accounted for in this way. At any rate, with the decline in the number of young operators, farmers as an occupational group are getting relatively old.

When the number of farms is projected for the next several decades, the situation of farmers of different ages must be considered separately. The peak number of farmers who were 35 years old or more had been reached in 1960; from then on, their numbers can be expected to decline in more or less the typical pattern of decline of past decades. These men will have entirely passed from the scene by 2000, but in the next two decades they will be nearly dominant in the number of farms. By 1960, the number of farmers born after 1925, who were thus less than 35 years of age at the time, had not reached a peak; their number will become important by 1980 and 2000 in the total number of farms. However, it appears that the peak number for this group will be much less than that for earlier age groups; but, once the peak for farmers born after 1925 has been reached, their decline is likely to follow the same general pattern as in the past. A third group is composed of those men and boys who had not entered farming by 1960 but will do so in the next few decades. In view of the generally unattractive income prospects for agriculture, especially for the smaller farms, it seems probable that the number of such farmers will be relatively small. In any case, they will not be an important part of the total until 1990, but will be a largely dominant part by 2000.

The result of these various changes by age groups is likely to be a decline in the number of farms to 2 million or less by 1980, and to 1 million or less by 2000. The exact figures at these future dates, of course, cannot be stated exactly. But both the direction and the general magnitude of the changes seem fairly clear. One consequence of these probable changes is that the abnormality of the age distribution will get worse, reaching its peak in about 1980, before it becomes more nearly normal.

In spite of these drastic changes, the future total area of land in farms will probably remain about as at present. During the past two to three decades, while the number of farms was shrinking rapidly, total land area within farms rose. A great deal of this rise took place in the Great Plains; there a larger area of land previously owned by non-residents and used on a more or less common basis was brought under the ownership or control of established farmers. This process cannot go much further, for most of this Plains land is now included in farms. Elsewhere, as the number of farms has shrunken greatly, the area within farms has sometimes increased slightly, sometimes remained about constant, and sometimes declined, but generally has not declined proportionately to the decrease in farm numbers. This is another way of saying that, by and large, most of the land from farms no longer operated as separate farms has been included within the remaining farms. Generally, we expect this process to continue. Some land not now in farms will be brought into farms, there will be a considerable rise in the area of land (often forested) not in farms, but on the whole a stability of total farm acreage seems probable.

If total farm area is to remain more or less stable while the total number of farms shrinks rapidly, then the size of farms will have to rise sharply. For a long time, the average acreage per farm was relatively constant, at 150 acres or less per average farm; it has risen in recent years to about 300 acres per farm. With the projected changes in the number of farms, it will rise to 1,100 acres or more per farm by 2000. This will be the average size; obviously there will be much variation about this average, by types of farms and among farms of the same type in the same general locality. Some farmers will manage to increase their dollar volume of ouput from present farm acreage by more intensive management; but, in other cases, land of low productivity now in crops will be shifted to pastures or forests, and thus be less intensively operated.

The projected change in the number of farms has many implications for small towns in rural areas, which in the past have provided services for farm families; for rural services of all kinds, which will face a drastically reduced clientele; and for the optimum road network and the many social services geared to road patterns. These are especially important factors in the total rural society and economy, but since they are less directly relevant to soil conservation, we shall not examine them further at this point.

Total agricultural output must rise in the future, as total population, and hence total demand for agricultural products, rises. A reasonable projection is that gross agricultural output will reach $42 billion by 1980 and $57 billion by 2000, compared with $29 billion in 1960 (all measured in 1960 prices).[9] With gross output doubling over the next generation, while farm numbers are falling by nearly three-fourths, some truly dramatic changes in output per farm are in prospect. As Figure 48 shows, the number of farms has fallen and will fall much further; total gross farm output has risen and will rise much further; and output per farm (in constant prices) increased nearly fourfold from

[9] Landsberg, Fischman, and Fisher, *op. cit.*

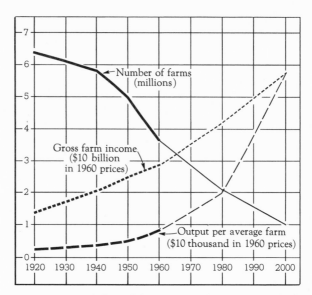

Figure 48. Number of farms, gross farm income, and
output per farm, 1920 to 1960, and projections to
2000.

1920 to 1960, and will be seven times the 1960 level by 2000. On
this basis, in the year 2000, the average farm will have a gross income
of $57,000. The business aspects of farming will loom larger than the
purely technical ones, in this situation, although techniques will be
more sophisticated than today. Moreover, this greatly larger farm, in
acreage and in output, will be operated with no more, perhaps less,
labor than the average farm of today. It will be a "family farm" in the
sense of being operated almost wholly by the labor of the farm operator and a little family labor; it will not be a "factory in the field" in the
sense of a large hired labor force.

This projected change in output per average farm has major implications for the distribution of farms by size. (See Figure 49 and
Table 25.) In 1939, two-thirds of all farms had a gross income of less
than $2,500; by 1959, this proportion was well under half; and we
project that it will be less than 10 per cent by 1980, and that, by 2000,
farms as small as this will have disappeared. Just as today farmers no
longer weave their own cloth from wool shorn from their own sheep,
nor make their own shoes from hides tanned from their own cattle, by
2000 the small, partly self-sufficient, essentially handcraft farm will
have become a museum piece. At the other end of the scale, farms with

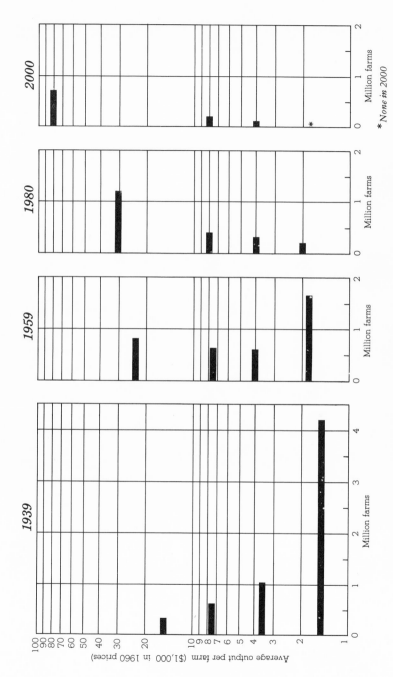

Figure 49. Number of farms by gross income class, 1939, 1959, and projected for 1980 and 2000.

TABLE 25. Estimated number of farms by gross income class, 1939, 1959, and projected for 1980 and 2000

(Prices in 1960 dollars)

Sales per farm	1939		1959		Projected–1980		Projected–2000	
	Est. mid-point of class	Num-ber of farms (million)	Est. mid-point of class	Num-ber of farms (million)	Est. mid-point of class	Num-ber of farms (million)	Est. mid-point of class	Num-ber of farms (million)
Under $2,500	$1,800	4.18	$2,000	1.64	$2,000	.2	$ —	none
$ 2,500–4,999	3,500	1.02	4,000	.62	4,000	.3	4,000	.1
$ 5,000–9,999	7,500	.58	7,500	.65	8,000	.4	8,000	.2
$10,000 and over	15,000	.31	23,000	.80	31,000	1.2	78,500	.7
Total		6.10		3.70		2.1		1.0

SOURCE: Jackson V. McElveen, ''Farm Numbers, Farm Size and Farm Income,'' *Journal of Farm Economics*, Vol. 45, No. 1 (February 1963), Table 1, p. 2.

a gross income of $10,000 and over accounted for only. 5 per cent of all farms in 1939, although they produced more than 30 per cent of total output; by 1959, they were nearly 22 per cent of the total number and produced nearly two-thirds of total farm output. According to our projection, such farms, by 1980, will be well over half of all farms, producing almost 90 per cent of total output; and, by 2000, they will be three-fourths of all farms, accounting for well over 95 per cent of all output. In all these comparisons, adjustments have been made to 1960 prices, so that the changes are real, not merely variations in prices. Moreover, the average size of the farms within each group, but most notably in the group of largest farms, has increased greatly. The average farm with a gross income of more than $10,000 had a $15,000 gross income in 1939 and $18,000 in 1959; and (as we project it) will have $31,000 in 1980 and $78,500 in 2000. These giant farms of the future, still "family farms" in the sense of the labor used, will surely have different attitudes toward, and abilities for, soil conservation than the smaller farms of today.

The cash costs of purchased inputs into farming, as noted in Chapter 5, have risen in proportion to gross farm income, especially in the years since World War II. In part, this has been the outcome of more advanced technology and of increased specialization of farming. Much that was once produced on the farm, or not used at all, is now purchased. Farms probably will grow even more specialized, purchasing still more of their inputs, relying still more on specialists to perform marketing and other services, with a consequent higher ratio of cash costs to gross farm income. As a result, the farms will be even more vulnerable than now to change in prices of purchased inputs and of commodities sold. Although of great importance to agriculture as a whole, this development is not so directly relevant to soil conservation and, hence, we shall not consider it further.

Changes in Land Use

Almost surely, net changes in land use during the next generation will be less than those which took place during the great pioneering period of American history, when forests were cleared and prairie plowed; but they may be equal to, or possibly greater than, those which have taken place during the past generation.[10]

The area used for various urban purposes and for recreation is likely to increase considerably on a relative scale, but these uses now take comparatively small areas of land. But the land required for agriculture can, and probably will, shrink somewhat, as rising crop yields more than outbalance the modestly growing demand for farm commodities. It is more likely that the area in forests and in grazing will shrink than increase, but shrinkage will be by comparatively small acreages. Other miscellaneous changes are likely, but do not invalidate these general conclusions. These are estimates of the future, and the actual fact may be different; but there is a rather high degree of consensus among professional workers, not only as to the direction of changes, but also as to relative magnitudes. A degree of stability in the total national pattern of land use has been achieved.

Gross shifts among various land uses, however, will be considerably larger than net shifts. The U.S. Department of Agriculture has estimated that 68 million acres will shift out of use as cropland between 1959 and 1980, but that 17 million acres will shift into crops; that 30 million acres will shift out of grassland pastures and ranges, to be more than offset by 48 million acres shifting into this use; that 32 million acres will shift out of forest, to be nearly offset by 27 million acres shifting into this use; and it estimates other partially compensating changes.[11] Even if the shifts in land use should not occur in exactly this fashion, it seems fairly certain that total shifts will loom much larger than net shifts.

From the point of view of soil conservation, it is entirely possible that the shifts of land use within farms will be the most important of

[10] Clawson, Held, and Stoddard, *op. cit.*
[11] *Land and Water Resources—A Policy Guide,* U.S. Department of Agriculture, May 1962 (slightly revised, September 1962), p. 43.

all. We know of no studies that have explicitly considered this problem and the possibilities for the future. We made the point in Chapter 5 that the Soil Conservation Service (SCS) has had to work with farms and farmers as it found them, but that both were changing rapidly. When the SCS began, the average size of farm was 150 acres, today it is over 300; and by 2000, it may well be over 1,100 acres. A vastly different group of land resources, as well as a greater amount, is thus typically included in an average farm. Shifts of land use to bring actual use more in line with land capability are now more nearly in the control of each farmer. The situation will differ greatly, in different regions and for different farms, of course. For example, in hilly or other situations with much steep or sloping land which is not well suited to crops but flat bottoms that are good, there could be drastic shifts within farms; crop acreage could be concentrated on the better lands and other land used for pasture or forests.

All of these changes will mean a much greater intensification of land use on the better lands—interpreting "better" in the sense of the SCS land-capability classification. The nearly level, deep, generally fertile, moderately textured lands which make up Class I usually will have far more fertilizer applied, be used in more intensive rotations such as continuous corn, and otherwise be used with greater intensity. The same will apply to Class II and III lands, but in diminishing degree. At the same time, cropping may well retreat from Class IV and poorer lands which will not respond well to increased inputs and will not lend themselves to the vastly greater scale of farm operations that will be necessary if the projected changes in output per farm are to be realized.

Attitudes toward Nature and Conservation

There is another range of possible changes in the future. These changes are more difficult to appraise, but possibly more important for soil conservation: the attitudes of people generally toward nature and conservation. On the one hand, more of this country's total population will live in cities, remote from the land, with not even secondary roots in farming. Even Grandpa will no longer live on the farm, where one can visit in summertime. On the other hand, the level of education is rising, and more people know far more about science, including biological science, than was true a generation or more ago. There is a common tendency to glorify the farm people of the past, and stress how close they lived to nature. In fact, many of them neither knew much about what they saw nor really cared about it. Many an alert schoolboy or Boy Scout today understands ecological relationships much better than did his pioneer great-grandfather. Schools and various conservation organizations are producing many well-informed persons in the urban populace who are directly interested in conservation generally. The vast majority of people are still ignorant of and indifferent to resource conservation problems; and the problems grow more complicated with the passage of time. But we hazard the judgment that public knowledge and concern about resources is rising in relation to the difficulties of the problems; and this, in spite of the rural-urban shift of population.

Perhaps we can conclude that conservation in the future will still be "good," in the sense that it is widely accepted as desirable, but that agricultural fundamentalism will have much less appeal. People who have always lived in cities, whose parents have always lived in cities, and whose grandparents now live there and perhaps always have, will be harder to impress with the homely virtues and necessity of rural living. They may be friendly to farmers, but they may also be less likely to concede special virtue to the man of the soil. They may be even more interested in, and better informed about, conservation of all kinds, including soil conservation, and at the same time more hardheaded in judging the national worth of specific programs.

We cannot, of course, be sure about these future attitudes; other factors than those we have considered may be important; or the results may be different from our estimates. But soil conservationists of all kinds would do well to ponder the possible changes in the next generation's attitudes. Soil conservation in the United States has been a public, especially a federal, program; its form, scope of operations, and perhaps even its continuance will depend increasingly upon the attitudes of nonfarm people.

Chapter 12

SOIL CONSERVATION IN
A CHANGING FUTURE

WHAT do the general social and economic trends of the future mean for soil conservation? By soil conservation, we mean the whole broad movement or program for the preservation of basic productive capacity in soils, forests, and grazing lands; nor do we exclude some efforts at building additional capacity. We also have in mind much more than the programs of the federal government, and more specifically those of the Soil Conservation Service (SCS), although these are a major part of the whole. The efforts of farmers and other landowners, of local and state governments, conservation organizations of all kinds, and others are included.

Presumably, everyone would generally approve of soil conservation in this broad sense. It is better to maintain, and perhaps to increase, basic productive capacity of soils, forests, and range lands, than it is to see them deteriorate. But we do not accept soil conservation as an overriding goal; it costs money, or requires giving up other alternatives, and has to compete with many other desirable ends, both public and private. We think it must be considered in the light of these costs and alternatives. Moreover, even if soil conservation is good, criteria of efficiency are still applicable to its attainment.

In retrospect, the soil conservation movement of the past generation can be viewed as primarily a social movement, one directly geared to the economic and social circumstances of its time. Soil erosion was an old problem, and an increasingly serious one, long before Hugh Hammond Bennett began to preach about it. Even his evangelism and drive might have come to nought a generation earlier. In the early thirties, the nation was ripe for social change; recognizing this, one can still give every credit and honor to Bennett for his shrewdness in taking advantage of those times. The President, the Congress, farmers, and the general public were willing to experiment with new programs on the land and with new social and governmental arrangements, and were ready to provide money to carry them out. Whereas there was far from complete agreement on all this, and Bennett was often in the forefront of controversy, all new social movements are similar in these respects.

If the soil conservation program as originally conceived and begun had not changed considerably over the years, that lack of change would almost surely provide evidence that the program today is out of step, for economic and social conditions have changed greatly. But program change alone is not necessarily proof that the changes have all been in the right direction or to the right degree. The program adjustments to date almost surely are not as complete as they would be some day if the general society and economy were to stop changing at this point. Soil conservation programs, like many other programs, often lag behind the rate of change.

Above all, the United States faces a dynamic and rapidly changing economy and society during the next generation. We have considered briefly some of the major changes that are most likely to affect soil conservation more or less directly in Chapter 11. The soil conservation movement must take them into account; it can change slowly and grudgingly in response to numerous outside forces, or it can take the initiative in adapting to that future—even, in some degree, actually helping to mold that future. One might argue that economic and social changes affecting land will work out more quickly and smoothly in the future than in the past; change begets change, and custom has surely been affected in rural America. But the scope of land use and other agricultural adjustments during the next generation will be great enough so that one can surely expect resistance and many lags.

Problems Facing the Soil Conservation Movement

What are the general problems which the soil conservation movement must face during the next generation? We think five are major:

1. A continuing excess capacity in agriculture.

Crop surpluses are a fact of life in today's agriculture, and indeed in the whole national economy. Agricultural programs are the most costly of the federal government's purely domestic programs. The actual surpluses are evident, but the potential ones are the more important. We have suggested problems involved in potential increases by mentioning the changes in output that would result from high guaranteed prices for agricultural commodities. Without very large investment and without additional manpower, total agricultural output in the United States could almost surely be doubled in a decade, were there any reason to do so. The consensus of professional opinion is that actual and potential surpluses will continue at least for twenty years or so; beyond that point, there is less general agreement. But we take the position that the march of science and technology will continue, perhaps even accelerate, its pace, and thus that the potential, if not the actual, surpluses will continue at least to the year 2000.

The whole public and private rationale for soil conservation must be different in a generation of abundance than in one of scarcity. The old cries of impending doom fall on deaf ears, if they do not provoke derision. The economic rationale of expenditures, whether private or public, to build or to maintain capacity in the face of present excess capacity, is different. We do not in the least go to the opposite extreme from the early dedicated conservationists such as Bennett, and say that no organized soil conservation program is necessary or desirable. But we do insist that the framework has changed.

2. *A continued rigidity of public farm programs, both federal and state, but a shrinking political power of agriculture generally.*

Bennett had to fight desperately for a place in the bureaucratic sun, and no small part of the history of soil conservation in the past generation is due to the ensuing bureaucratic and organizational fights. But, over the years, the Soil Conservation Service and the soil conservation districts have established a well-defined place in the governmental and organizational picture. So have such agricultural programs as the Agricultural Conservation Program (ACP), the Extension Service, and others. Each has a specialized role and together they encompass a wide gamut of specific measures. There may be considerable doubt that the sum of the various parts is equal to a comprehensive approach, or is even what any one of the component elements would try to do were it given the entire responsibility for the whole agricultural program. But one can be fairly sure that changes will come slowly in the future, incrementally, and rarely at the cost of any serious weakening of any entrenched interest—whether a bureaucratic or private interest. If Bennett were alive today and attempting to organize new ways of meeting the evolving social and economic needs of the times, we think he would have a much harder task than the one he faced in the early thirties.

At the same time, the political power of agriculture as a whole is shrinking rapidly. In part, this is caused by the shift of population from farms and small rural towns to the small and larger cities. But it is a result, in important part, of recent court decisions which have forced the redistricting of congressional districts and state legislative election districts within states to more nearly equalize the population for all districts. It has been estimated that in 1962 twelve congressional seats shifted from rural to urban constituencies as a result of adjustment among states brought about by population changes measured by the 1960 Census. The days of a powerful farm bloc in Congress are gone; even if the farm bloc could agree—which it has not been able to do in recent years—it can pass legislation only with the consent of urban political forces. Perhaps as important as the loss in the number of seats is the fact that able members of the Senate and the House of Representatives increasingly are becoming uninterested in serving on agricultural committees. The congressman from a safe rural district, who by sheer survival has been elevated into a committee chairmanship, will be a less important factor in the national political structure in the future. The same situation, in general, will exist in state capitols, although state programs have been less important for agriculture than have federal ones. Drastic shifts in political power, from rural areas to cities, took place in the 1964 election in Michigan and in some other states; but greater shifts within states will occur in 1966, 1968, and later elections. Astute politicians will still make their obeisance to rural virtues, and will help provide or develop programs to aid rural constituents as well as others; but, when contests arise, farmers simply will not have the political strength in the future that they have had in the past.

3. *A drastically shrinking number of farms, and a consequent enlargement of the size of the average farm.*

The evidence for this reduction in the number of farms and estimates of its magnitude have been presented. As we have suggested before, the best conservation use of land in one farm of 1,100 acres may be quite different from the best use of the same land when it is in four farms of nearly 300 acres each. New plans will have to be made for farms, perhaps several times; and, to some extent, the plans will have to be based on different ideas and concepts. If the soil conservation movement is to contribute constructively to the changing agricultural society and economy, it must not become an obstacle to changes in the number and size of farms, but rather a means of helping guide and direct such changes into a more rational rural structure and society. Special problems of rural leadership and of rural clientele will arise for soil conservation as for other agricultural programs.

4. *Changes of land use within farms, as well as changes involving whole farms.*

With larger areas of land, often with new combinations of land capability, the farms of the future can make very different use of land than in the past. Past changes in land use seem to have been in the direction of better adaptability of use to capability, but they have been slow. A major obstacle has been the rigidity in the size and boundaries of farms, and the desire of each farmer to remain on his land; the problem of conservation had to be solved within a very restricted geographic unit. If farm size quadruples over the next generation, as we estimate that it will, then the geographical size of the planning unit increases correspondingly and many new possibilities open up.

In this connection, we need to re-emphasize the very large discrepancy between land capability and land use. Some opportunities for adjustment of land use exist within farms, and will be present especially within the larger farms expected for the future; but vastly greater opportunities exist within localities. Almost every study which we have been able to find reveals a substantial discrepancy between land use and land capability within the locality or county. Whereas there is need for major adjustments toward better land use within both the states and broad regions, important gains also can be made locally.

5. *The frequent lack of profitability to both private and public investors of soil conservation measures, and a general over-capitalization of agriculture.*

Since land use situations differ enormously, one must be careful about generalizations. In some situations, the investment of private funds in soil conservation is profitable over the time period which the farmer is willing to include in his calculations. Frequently, however, profitability depends upon an ability or willingness to adopt a more intensive and higher level of management of land and other inputs—something not always possible for many farmers and other landowners. Technical services or funds from public sources may make programs attractive which otherwise would not be, and may be justified because of general social benefits not available to the individual. Nevertheless, when allowance is made for these factors, and perhaps other factors and situations, the fact remains that frequently, perhaps generally, soil conservation measures are not highly profitable. In a great many situations, someone would have to spend more money than society is willing to count as a benefit in order to reduce soil loss to the level soil conservationists would consider desirable or even tolerable.

Two other facts contribute to this lack of profitability. One is the general excess capacity in agriculture. When there exists already more productive capacity than the nation currently needs, and when it appears that a combination of other forces will increase that capacity as fast as, or faster than, new capacity is needed, then the wisdom of large-scale investment—whether public or private—in soil conservation is definitely diminished. The other fact is the general over-capitalization of agriculture. In considering additional investment for soil conservation, one should rationally weigh the additional output in relation to the added cost—the marginal revenue as against the marginal cost. At least in many situations, the present high investment and low returns from farming certainly discourage further investment.

Broad Dimensions of an Ideal Soil Conservation Program

If the foregoing constitute at least some of the major problems which the soil conservation movement will face over the next generation, what are the general dimensions of an ideal soil conservation program? Some of what we put forward as goals may be unattainable, at least quickly and easily; and all are necessarily presented in general terms which would have to be refined and spelled out before they could be implemented. But a general goal, or set of goals, can provide a useful guide to actions in the short run. To this end, we suggest the following major components in a soil conservation program:

1. *A nationally accepted, clearly defined set of objectives for soil conservation, especially for public programs in this field.*

"Conservation" means many things and, therefore, little that is specific. We have noted how the content of the federal soil conservation program has shifted greatly over the past thirty years or more, usually shifting without new legislation to define the new objectives.

We have also noted the inconsistency between output-increasing and output-decreasing federal programs, many within the U.S. Department of Agriculture. One can be critical of these conflicts and of gradual changes in the content of programs, which often seemed induced by a desire to win political support. But the President and the Congress have never given a clear policy direction to federal programs in soil conservation. Should the programs seek to maintain basic productive capacity in the country's soils, but no more? Or should the programs seek to build additional basic productive capacity? If the latter, to what extent? Should the productive capacity be added everywhere, or only on some lands, with offsetting retirement or shift in the use of other less suitable land? Should the programs seek to stimulate farm inputs beyond the point which farmers would find profitable under the stimulus of market prices for their products? These are the kinds of questions on which a clear policy position is needed.

If there were a clear policy directive on these and related problems, then it would be possible to judge the progress, or lack of it, in attainment of the policy goals; to judge the rationality of specific programs aimed at these goals; and to evaluate the optimum amount of public investment in various programs. Unfortunately, the chances of formulating and implementing such broad policy goals for soil conservation do not seem good. A consensus in favor of national programs is possible, largely because goals are not specific; conservation is more likely to mean all things to all men—and thus gain support—when it is vague than when its meaning is precise. The advocates of public soil conservation programs often seem unwilling to take their chances on the measurement of specific programs against clear goals. But, in the absence of clear goals, drift or imprecision is inevitable.

2. Dependable, useful, informative data and analyses of the accomplishments and remaining problems of soil conservation.

The lack of relevant information has been stressed at numerous points in this book—it might almost be called our theme, if the book has a single theme. There are many figures, but crucial ones are lacking; there have been surveys (notably the National Inventory of Soil and Water Conservation Needs), but their results have not been adequately analyzed and presented. There are hints, but direct answers to the basic questions are often missing. As we have pointed out, the available data permit an estimate that the soil conservation job as it appeared in the early thirties had been 20 per cent completed by 1958, or had been 60 per cent completed—depending upon which of equally reasonable assumptions are used in analyzing the available data. Faced with urgent soil conservation problems in the past, neither federal agencies nor farmers have been much concerned with critical analysis. That attitude is not only understandable for the earlier years, but perhaps can be justified; but the future will require, we think, a very different approach.

As a specific suggestion, we propose that some small part (possibly no more than one-half of 1 per cent) of the total federal soil conservation funds be earmarked for an initial survey or inventory of the soil conservation situation in the United States today; and that similar sums be made available annually for updating, extending, and refining the initial survey. Such a collection and analysis of data should be as objective as it can be made, not normative in the sense that the Conservation Needs Inventory was. The nation should know just how much land, of what qualities and in which locations, is suffering from carefully defined degrees and kinds of soil erosion; how much land, of what qualities, and in which locations has had specific and identifiable conservation measures carried out in the past and how effective they were; and how much land, of what qualities, and in which locations is physically capable of reacting to carefully defined types of improvement, such as drainage, irrigation, terracing, reseeding, and the like. The full details of such data collection and analysis would have to be worked out, but the emphasis should be upon accuracy and objectivity. Programs of soil conservation could be built upon such a data base, but should not be built into the process of data collection and analysis. In view of the sums of money annually spent for soil conservation and the relatively long period during which such a program has been carried on, it is high time more is known about what is being done.

3. *Reasonable consistency of soil conservation programs with other agricultural and resource programs.*

The continued basic inconsistency of one group of public programs for agriculture with the objective of increased basic productive capacity, while other and more expensive programs sought to protect agriculture from the effects of excessive output, has been noted. This conflict is not restricted to the program of the Soil Conservation Service and the Agricultural Conservation Program; it extends to other programs, including research, in the U.S. Department of Agriculture, and to other federal programs. Federal officials, from the President to the

Secretary of Agriculture to program directors, over the years have tended to minimize or deny such conflict, yet the record unequivocally reveals it. In practice, the preservation of basic productive capacity in land quickly and naturally shades into building additional basic productive capacity, and frequently includes stimulus toward greater annual inputs of productive factors as well. This extension of the purpose of conservation practices made sense to the individual farmer, whose output was too small to have any significant effect on prices of agricultural commodities; and it made sense to the SCS and ACP, each striving to build popular and political support for its whole range of programs. But its wisdom for the nation was clearly less.

Ideally, for the future, these conflicting purposes of agricultural programs will be resolved. Individual farmers, of course, are free to increase basic productive capacity whenever this appears profitable to them; but should they be given public subsidy to help them do so? Moreover, one should not rule out all increases in basic productive capacity. It is at least theoretically possible to build new capacity in some situations, and eliminate or reduce existing capacity in others, and yet to leave both the nation and all the individuals concerned better off. It is possible, but not generally common, perhaps. The problems of adjustment are considered below. As far as soil conservation is concerned, shifting the use of land to less intensive uses may often be the cheapest soil conservation from a national viewpoint. Under many circumstances today, the net value of additional agricultural output is negative from a national viewpoint.

4. *More fluidity in the adjustment of agricultural inputs, particularly of labor from the farmer and his family.*

One major factor responsible for the persistent tendency of agricultural output to rise, as we have pointed out, is the unwillingness, or inability, of farmers to withdraw from agriculture, even in the face of very low incomes. There are powerful basic reasons underlying such rigidity of labor input into agriculture. We think it neither possible nor desirable to make drastic and immediate shifts of manpower out of farming. But many men are shifting out—rather, many are not moving in, as young men refuse to try to operate substandard farms having long-run income prospects which are so poor. Perhaps the single most dramatic and important change in American agriculture in the past generation has been the halving of the number of farms; but this has been carried out with almost no public help or guidance. We think it would be possible to provide some help under such circumstances. It would be possible to reduce the financial and social costs to those who move; to ease the burden of those who cannot move; to increase to some extent the rate of movement; to improve land use for those who remain; and, above all, to improve the social environment in which farmers live and work. Programs of farm consolidation and rural community reorganization begun in the next few years could drastically improve the situation of the next generation.

Programs to aid in developing the best farm and community organization, as the number of farms shrinks drastically and the size of farms rises correspondingly, will involve far more than soil conservation agencies. But the soil conservation programs, ideally, should facilitate rather than impede these inevitable adjustments; and they have both a great responsibility and a great opportunity to bring about a much better use of land.

5. *Critical examination of the economics of soil conservation and the provision of less generous public subsidies where the results are doubtful.*

Strict cost accounting and strict economic profitability have never governed this country's soil conservation programs as a whole, and we doubt if they will; but economic considerations might well loom larger in the future than in the past. This applies not only to programs which create additional basic productive capacity; but also to the ways in which existing capacity is maintained. In our opinion, several programs seem to have been costly in terms of results, or the physical results were not those which the nation really needed. We recognize that public, usually federal, subsidy is pervasive throughout agriculture, and also in other resource programs, such as flood control and navigation improvement—indeed throughout much of the American economy and society. Even if it were desirable, elimination of all this subsidy is not at all likely. But this does not mean that more critical analysis of the amounts, purposes, and recipients of subsidy would not be socially desirable. The economics of private investment in soil conservation probably deserves a closer look too, and public agencies perhaps should be more hesitant to recommend programs and practices when profitability is not certain.

6. *A better gearing of soil conservation programs to over-all resource management programs.*

Some important steps have been taken to make soil conservation programs more complementary to other programs, as emphasis has shifted toward watershed management, flood protection, and the provision of recreation opportunity. But these have been more a matter of the established soil conservation programs catching on to popular bandwagons than of any fundamental broadening of soil conservation as such. The latter is still viewed as basically a program for cropland and for farmers, with some attention to grazing land and forests. Soil conservation districts are largely governed by farmer-supervisors. The broad economic and social changes expected for the future will bring other resource uses and other users to the fore. The programs directed specifically at soil conservation should be re-evaluated and modified constantly to meet these changing social and economic conditions.

Forces and Processes

Will the problems be met and the objectives of an ideal soil conservation program be achieved over the next generation? What is a practical goal of accomplishment, in view of all the problems? What new, and now unanticipated, developments will occur to change the whole picture?

Answering such questions, important though they are, would plunge us directly into forecasting, as we defined the term at the beginning of Chapter 11. That is, one would have to weigh each of the factors we have mentioned, include any others that he could think of, and somehow combine them into a forecast of probable action. This we do not intend to do; it is too distracting; merely making it might divert attention from the more fundamental processes which need to be considered. Nor do we think it possible to make a meaningful projection, in the sense we used the term earlier. Any good projection must be sufficient, in the mathematical sense; that is, the assumptions should be sufficient so that the projection flows directly and inescapably from them, with only one possible answer. We feel that the situation as to soil conservation is too complex, with too many variables, to make a sufficient number of assumptions.

We can consider, and we think with profit, the forces and processes at work now and probably over the next generation which will affect, if not determine, the nature of the future soil conservation program. These perhaps can best be considered in terms of the actors.

Farmers and the private farm organizations will surely have a major effect upon soil conservation. They will adopt and carry out those programs which they think will be profitable—including the use of any public subsidies that may be available. As in the past, their main interest will not be in the conservation of soil as such, but in conservation as it affects present or future net income. The coming generation of farmers will be vastly better informed than their forerunners about the processes and hazards of soil erosion and about the methods of soil conservation. While exhortation and encouragement may still be necessary, future farmers generally will be capable of making independent judgments about soil conservation programs on their own farms. They will need dependable and applicable information, however, and they often will look to public agencies for it.

The reluctance of young men to enter farming, especially on low-income farms, the gradual dying off of the older farmers, and the unbalanced age distribution among farmers in many areas will present serious handicaps to soil conservation programs. A large number of farmers will have short planning horizons; they will not expect to continue personally for many more years, their sons will not be interested in the farm, and the land market will not reward them or their heirs for soil conservation investments. The best ultimate use of their farms, once the present operators can no longer continue, may be as parts of larger farms persisting in the same locality. In a great many situations, it may be difficult to enlist able, imaginative, progressive, and, particularly, young leadership for soil conservation as well as for other agricultural group activities. Many soil conservation districts are likely to be faced with the question of whether their continued existence as a separate district, as compared with consolidation with neighboring districts, is the best course of action.

The private farm organizations, whether for general agricultural purposes, for co-operative marketing, for credit, or for soil conservation, will face some of these same problems. Once created, such organizations to some extent provide their own excuse for being. Their employees, and even more their officers and leaders, have a vested interest in continued existence. Many of these organizations will face difficult problems of their own in coming years. On the one hand, they will have little time or energy to help with soil conservation; on the other hand, they may be looking for allies and support. For this, the soil conservation districts and their organizations are in a special position.

There are many rural landowners other than farmers, and there will be still more of them in the future. The Forest Service has estimated that in addition to farmers there are 1.1 million owners of small forests.[1] As cropping is stopped on some poor farm land and as more urban people can afford and will want places in the country (primarily for recreation), the number of such nonfarm rural landholdings is likely to increase. We know very little about such landowners or their motivations. We can be fairly sure that they do not own and operate their land in the expectation of making a profit from it. By and large, such landowners have not been closely drawn so far into soil conservation operations and we do not know much about their attitudes toward conservation. Like the general public, they probably are favorably disposed, ideologically, toward soil conservation—are likely to think that it is "good." But they probably will not react to the same appeals and approaches which have been developed for working with ordinary farmers.

The role of the land-grant colleges will surely change over the next generation. Through experiment stations, instruction, and extension courses these colleges have tooled up to serve agriculture; they may have now or in the future much excess capacity for their job. Some are already turning heavily to natural resource matters generally, not merely to agriculture; and some are working actively abroad in various forms of economic and technical assistance to the less developed countries. Their solid basis of public and political support is within agriculture, and they may be expected to guard that carefully. However, they will be affected by some of the problems we have outlined above. They have already experienced declining total enrollment in their agriculture courses, at a time when college enrollments generally have been mounting rapidly. The established patterns of activity in such institutions are difficult to change. They are likely to find soil conservation at least as popular among farmers as in the past, and perhaps more so; and this might become one of the lines of activity which would appeal to the nonfarm rural landowner and to the general public. The land-grant colleges and universities could be a major force, perhaps the dominant force, in achieving something like an ideal program of soil conservation. But we cannot be too sanguine; their specialized and established role is part of the big problem caused by the rigidity of agriculture's organization and program.

[1] *Timber Resources for America's Future,* Forest Resource Report 14, Forest Service, U.S. Department of Agriculture, January 1958.

The U.S. Department of Agriculture as a whole has many points of similarity with the land-grant institutions—similarity in the past, now, and probably in the future. It, too, is tooled up relatively well for its job—perhaps too generously so, compared with many other public programs and public needs. As with any other agency, there are many entrenched interests and many rigidities. For example, one might argue that the whole ACP program should be liquidated; but anyone even moderately familiar with agriculture can imagine the uproar that this proposal would create, not only in the Department of Agriculture, but in the Congress, among farmers and farm organizations, and perhaps above all among the suppliers of agricultural inputs. The role of the limestone producers association has already been mentioned as an example. But it is upon the Department of Agriculture, and more specifically upon the Secretary of Agriculture, that the responsibility falls for incompatible and inherently conflicting programs—programs to build more productive capacity carried on at the same time as other and more expensive programs for the management of agricultural surpluses. One of the Secretary's feet is on the gas, the other on the brake—though neither of his feet may be alone on either pedal. In saying this, we fully recognize that no Secretary of Agriculture is a free agent in this matter; it has been neither ignorance, perversity, nor indifference that has caused a succession of secretaries to accept or tolerate the discrepancy. It is also incumbent upon the Department, and again specifically upon the Secretary, to determine how much to spend for soil conservation, and under what terms of subsidy to landowners, especially as compared with other agricultural programs. Again, no Secretary is a wholly free agent.

It is when one turns to the Bureau of the Budget and to the President that the chances for major change in future federal soil conservation programs perhaps become greater. The diminishing political power of agriculture will be a steady force toward smaller public subsidies, toward more efficiency in the use of such money as is made available, toward a demand for a better analysis of where agriculture is and where it is going. We do not suggest that any future President or the Bureau of the Budget will be hostile to soil conservation generally. But other programs are likely to loom larger in the thinking of each, and each is likely to be less impressed with political pressures from agricultural interests. For example, urban demands for federal financial help in slum clearance and in housing will have far more political voltage. The President and the Bureau of the Budget cannot be instruments for specific change within soil conservation—they are too remote and too preoccupied elsewhere for that; but they could apply pressure where it would be effective.

Some of these same considerations enter when the future role of the Congress in soil conservation is considered. There will be far fewer congressmen in the future with predominantly farm, or even rural, constituencies; there will be fewer political careers to be built upon representation of farm interests. We have noted how able congressmen are already tending to leave the congressional agriculture committees. Farmers, their organizations, and supporters generally will be less able to influence the action of congressmen directly, or through them to influence the action of the President, the Department of Agriculture, and other parts of the Executive Branch. To the extent that soil conservation as a movement develops real public support among other rural landowners, it will have some political strength not limited to agriculture.

In summary, from the point of view of federal action, the President and the Bureau of the Budget are likely to become the motivating force for changes in agricultural and soil conservation programs. Increasingly, the Congress will be less interested, and less able when interested, to retard or stop changes which agricultural interests in the past would have opposed successfully. And the Secretary of Agriculture is likely to get more of his real directives from outside of agriculture. These changes will come gradually and will be general tendencies, observable over a decade or two or three, not easily pinpointed as to specific dates. They need not be hostile to soil conservation; we simply are saying that the center of decision making and the basic considerations for it are going to shift substantially over the next generation.

The attitude of the urban population in the next generation is likely to be important for soil conservation. On the one hand, there is the steady increase in the number of urban people who have no contact with farming, no agricultural fundamentalism, and are likely to be skeptical of claims for special rural virtues. On the other hand, many of these people are becoming better educated on nature as such, and often are more concerned about resource problems. They probably will be favorably disposed toward soil conservation, but also critical of it, and will not be committed to past programs and entrenched positions. It is unlikely that they could change the soil conservation programs or would try directly to do so. But their attitudes, as reflected to their political representatives, might well determine how much general governmental support would be given to soil conservation over the next generation.

We have purposely saved for the last our consideration of the soil conservation districts, the National Association of Soil Conservation Districts (NASCD), and the Soil Conservation Service. These are the organizations and interests which will be most directly concerned with soil conservation over the next generation. While they alone cannot make all the changes that are most needed, they must surely take a major share of the responsibility for change. Change could be thrust upon them, rammed down their throats, but that would be a difficult and inefficient way for the nation to bring about change; and it might destroy these specialized soil conservation agencies in the process.

In the past, the SCS, the soil conservation districts, and the NASCD have been concerned primarily, almost exclusively, with agriculture, and more specifically with cropland; they have dealt almost entirely with farmers, and with a rather selected part of the total farming population; and their chief, if not sole, concern has been to preserve and to build basic productivity in the soil. We do not intend the foregoing statement as criticism, but as description. These groups could well argue that their job, as defined in the past, was to deal with cropland, to maintain or to build its productivity, and thus to work with those farmers—often operating the larger and more commercial farms—who were interested in soil conservation. All three of these groups have been successful, at least to some degree. They have become established in a political or government sense—some might say have become encrusted in the process. In fighting for its existence, the SCS has won to the point where the most unsympathetic future administration would hardly undertake to liquidate the program or the agency, or even to curtail it critically. Nearly the whole nation is now organized into soil conservation districts, though some are surely weak as independent entities; in this sense, the district idea has won acceptance. In fact, the existence of established districts may well be a major obstacle to change in the future. And NASCD also is an established national organization.

For the future, all three groups must pay increasing attention to all natural resources, not merely to cropland; must increasingly deal with all landowners, not merely part of the total farmer population; must be increasingly concerned with all land uses, not merely with agriculture and cropping; and must increasingly depend upon careful and critical analysis, and less upon exhortation. In our judgment, failure to change in these directions would result in the gradual exclusion of each of these three groups from natural resource matters generally, and to some extent even from soil conservation, as they have defined conservation matters in the past. In other words, the three must almost surely assume a larger role or be forced into a smaller one—they must "advance or wither."

We recognize that the specialized soil conservation organizations— SCS, the districts, NASCD—alone cannot solve the major problems facing agriculture. Alone, they cannot guide, or have much influence on the reduction in the number of farms, with all the shifts in human population which this will entail; alone, they cannot direct or control the farm consolidation and the resulting changes in land use, which are implicit in the changes in the number of farms; alone, they cannot bring about the reorganization of rural communities which a drastically reduced number of farmers and of persons in small towns will demand. If any one, or any combination, of these three specialized soil conservation organizations undertook alone to perform these basic jobs in agriculture for the next generation, it would arouse a torrent of criticism and opposition from other agencies and organizations, public and private, which consider some of these activities as their job. While the specialized soil conservation organizations cannot alone do these jobs, they are directly and immediately affected by how, or whether, the jobs are done. How can this dilemma be resolved?

Have the Soil Conservation Service, the soil conservation districts, and the NASCD achieved sufficient political maturity and strength so that they, alone or in combination, can seek to lead—perhaps not to lead but surely to stimulate—a combined or consortium attack, with other public agencies and private organizations, on these major agricultural and resource problems which exceed the grasp of any single line of approach? Are these three specialized soil conservation organizations prepared to make adjustments, perhaps to give up some of their historic roles as well as to take on some new ones, to face these future problems? Are they prepared to participate fully in a comprehensive program of agricultural adjustment and progress which some future far-sighted Secretary of Agriculture may inaugurate? And if they are prepared to try, how successful can they be in achieving these objectives?

APPENDIX TABLE 1. Conservation practices newly applied in soil conservation districts, 1941–1963[a]

Practices	Unit	1963	1962	1961	1960	1959	1958	1957	1956	1955	1954	1953	1952
Contour farming	1,000 acres	6,105	4,789	3,239	2,488	2,070	2,297	2,660	2,966	3,147	4,127	3,342	2,911
Cover cropping	1,000 acres	4,985	5,599	4,325	3,833	3,116	4,145	4,110	3,832	4,507	4,839	3,652	3,296
Strip cropping	1,000 acres	667	764	659	666	728	769	896	971	948	930	791	805
Seeding ranges and pastures	1,000 acres	2,693	2,899	3,145	4,157	4,761	4,279	3,431	3,278	3,135	3,008	2,093	2,920
Tree planting	1,000 acres	375	586	708	985	1,047	752	472	351	316	270	248	155
Pond construction	Number	53,886	57,301	55,670	58,429	66,809	63,060	81,178	78,869	86,574	64,521	60,497	61,578
Terracing	Miles	40,557	43,347	42,852	38,842	46,011	40,986	50,937	52,187	45,806	47,791	51,064	66,853
Diversion construction	Miles	3,310	3,562	4,079	3,911	4,972	4,878	6,000	5,590	6,902	4,808	5,682	5,526
Drainage	1,000 acres	(by miles)	1,829	1,671	1,490	1,757	1,443	1,301	1,510	1,554	1,578	1,560	1,315
Irrigation and leveling	1,000 acres	511	447	394	383	527	460	581	575	494	375	418	408
Improved water application	1,000 acres	2,409	1,333	2,167	1,705	1,166	1,291	1,412	1,720	1,714	950	784	538

Practices	Unit	1951	1950	1949	1948	1947	1946	1945	1944	1943	1942	1941
Contour farming	1,000 acres	3,003	3,377	3,537	3,171	3,394	3,275	2,575	2,848	1,496	1,166	690
Cover cropping	1,000 acres	2,876	2,963	3,013	2,550	2,619	2,370	490	180	571	n.a.	n.a.
Strip cropping	1,000 acres	838	849	775	690	826	601	475	588	591	558	337
Seeding ranges and pastures	1,000 acres	2,572	2,197	1,734	1,216	1,872	1,647	888	807	732	194	235
Tree planting	1,000 acres	162	153	113	93	56	36	33	26	29	84	49
Pond construction	Number	52,671	36,169	35,633	28,568	31,519	27,486	15,547	15,254	3,228	2,613	1,613
Terracing	Miles	72,685	90,882	101,895	89,245	101,363	99,482	71,910	53,777	44,149	30,195	21,146
Diversion construction	Miles	5,534	5,536	5,533	5,303	4,194	3,371	1,627	932	1,754	896	n.a.
Drainage	1,000 acres	1,259	1,129	1,227	1,017	1,103	753	300	203	83	27	1
Irrigation and leveling	1,000 acres	411	351	270	251	189	136	80	63	16	—	n.a.
Improved water application	1,000 acres	599	673	594	534	511	381	160	189	9	30	5

n.a.—not available.

[a] Data for 1953 through 1963 are based on fiscal years; before 1953, on calendar years.

SOURCES: Data from U.S. Department of Agriculture, Agricultural Statistics, 1946 (Washington: Government Printing Office), Table 712, and from the same publication for the following years: 1947, Table 731; 1948, Table 761; 1952, Table 777; 1962, Table 767; and from a 1963 press release.

SUBJECT INDEX

ACP (*see* Agricultural Conservation Program)

Acreage allotments, 107

Agricultural Adjustment Act, 44, 176, 284

Agricultural Adjustment Administration, 50, 60

Agricultural Board, 15

Agricultural Conservation Program: accomplishments, 179–83, 185–88, 192, 207, 226, 286(f)–87(f); acreage, 184; administration, 45, 51, 289; bureaucratic struggles, 50; cost, 87(t), 191; establishment, 60, 175–76, 316; example (1961), 178–80; future, 327; inconsistency in, 73, 182–83, 285, 287, 321; location, 189–91; objectives, 65, 176–77, 191–92; payments under, 178–83, 185, 188–192, 195–97, 270–71

Agricultural Extension Service: establishment, 39–40, 316; program, 47–48, 230; SCS conflicts, 45, 50, 68–69

Agricultural ladder, 305

Agricultural Stabilization and Conservation Service (*see also* ACP), vii, 50–51, 60, 152–53, 277

Agriculture (*see also* Crops, Farms, Land): availability of land, 4(f); capital in, 26, 318–19; changes (*see also* Land use, shifts), 8, 91–112, 257; glorification, 33; goals, 8, 92; income (*see* Income); inputs (*see* Inputs); output (*see* Output); political power, 316–17; population, 91–95, 298–99; prices (*see* Prices); productivity, (*see* Productivity); surpluses, 110–12, 241, 315–16; as use of solar energy, 24

Agriculture, U.S. Department of (*see* Department of Agriculture, U.S.)

Alabama, 187, 193, 216

Alaska, 130, 174

American Farm Bureau Federation, 50

American Forestry Association, 54

Anderson, Clinton P., 50, 51

Appropriations Committee (*see* Congress)

Arizona, 6, 174

Arkansas, 187, 248

Army Corps of Engineers (*see* Corps of Engineers, Army)

Attitudes toward conservation: farmers, 254–62; public, 230–32, 312–13

Audubon Society of America, 54

Benefits: from conservation, 13, 20–21, 26, 262–73; disassociation of, with costs, 21, 274–76

Bennett, Hugh Hammond: as crusader, 55, 58, 316; erosion concern, 60–64 *passim*, 230, 254, 314; food supply concern, 111; Reconnaissance Erosion Survey and, 157–58; role in early conservation movement, 41–47; Small Watershed Program and, 75–76; whole farm plan and, 67

Benson, Ezra Taft, 50

Brannan, Charles F., 50, 51, 52

Bureau of Agricultural Economics, 75

Bureau of Agricultural Engineering, 64

Bureau of the Budget, 327–28

Bureau of the Census, 94; Census of Agriculture, 192–202 (*see also* Author Index)

Bureau of Chemistry and Soils, 52

333

AUTHOR INDEX

Aandahl, Andrew R., 261n, 272
Aigner, F. D., 107
Aires, Ronald O., 292n
Allen, Carl W., 271, 271n
Anderson, H. O., 258–59, 259n, 270, 270n
Anderson, P. O., 218n
Arnold, C. J., 107n
Atkins, S. W., 266, 267n

Ball, A. Gordon, 271n
Barlowe, Raleigh, 16–19(f)
Barnes, C. P., 165n
Barnett, Harold J., 3–6, 29n, 32
Bauder, Ward W., 257, 257n
Baumann, Ross V., 261, 261n, 271n, 272, 281
Beal, George M., 257, 257n
Benedict, Murray R., 45n, 176, 176n
Bennett, Hugh Hammond, 58n, 63n, 66, 70
Blase, Melvin G., 217n, 261n, 273, 282
Bohlen, Joe M., 257, 257n
Bonnen, James T., 300n, 304
Boykin, Calvin C., Jr., 279, 279n
Bunce, Arthur C., 16n
Burch, Thomas A., 257, 257n

Capp, James H., 257, 257n
Carreker, John R., 267, 267n
Census of Agriculture, 94, 142, 143, 153, 193(f), 194(f), 198(f), 199(f), 201(f), 204, 212, 288, 290, 291
Christensen, Raymond P., 292n
Christiansen, R. A., 219n, 270n

Clawson, Marion, 94n, 95, 95n, 228n, 241n, 249n, 299n, 302n, 305n, 306, 311n
Cotner, Melvin L., 182, 182n
Coutou, A. J., 209(f), 209n, 210(t), 212n, 213n, 265, 265n, 268n

Dabney, Charles W., 36n
Dana, Samuel T., 35n
Dean, Gerald W., 271n, 273
Department of Agriculture, U.S., 21n, 63, 63n, 67–68, 68n, 70, 71n, 74n, 75n, 81n, 82(f), 83(f), 84n, 86(t), 87(t), 97(f), 100(f), 101(f), 117(f), 122n, 126–27(t), 132n, 134(t), 136(f), 137(f), 145(t), 146(f), 148(t), 155(t), 158n, 160(f), 161(f), 169–70(t), 172–73(t), 185n, 186(t), 202n, 205(f), 212(t), 223n, 238n, 239(t), 240n, 243(t), 245n, 246(t), 249n, 286(f), 287(f), 292(f), 293(f), 311n, 331(t)
Dietz, Martha A., 223(t)

Fischer, Lloyd K., 261n, 282
Fischman, Leonard L., 298n, 307n
Fisher, Joseph L., 298n, 307n
Frey, John C., 261n, 280–81

Gibson, W. L., 107n
Goodrich, Carter, 121(f)
Griffith, Ernest S., 29n
Grubb, Herbert Warren, 215, 215n, 216, 268, 268n

343

SOIL CONSERVATION IN PERSPECTIVE

by R. Burnell Held and Marion Clawson

designer: Edward D. King

typesetter: Monotype Composition Company

typefaces: Caledonia, Perpetua

printer: Universal Lithographers, Inc.

paper: Perkins and Squier SM

binder: Moore and Co.